Acknowledgements

The valuable assistance of the Organising Committee
and Panel of Referees is gratefully acknowledged.

Organising Committee

H.S.Stephens (chairman)	BHRA Fluid Engineering, U.K.
I.D.Fantom (secretary)	BHRA Fluid Engineering, U.K.
R.A.Bird	E.T.S.U., U.K.
O.De Vries	National Aerospace Laboratory, Netherlands
L.V.Divone	E.R.D.A., U.S.A.
R.I.Harris	Cranfield Institute of Technology, U.K.
S.Hugosson	National Board for Energy Source Development, Sweden
N.H.Lipman	Science Research Council, U.K.
O.Ljungstrom	National Board for Energy Source Development, Sweden
M.Pedersen	Technical University of Denmark, Denmark
G.G.Piepers	E.C.N., Netherlands
H.Selzer	ERNO Raumfahrttechnik GmbH, F.R.G.
J.S.Shapiro	Wind Energy Supply Co., U.K.
R.J.Templin	N.R.C., Canada
D.F.Warne	Electrical Research Association, U.K.

Citation

When citing papers from this volume the following reference
should be used:-
Title, Author, Proc. 2nd International Symposium on
Wind Energy Systems, BHRA Fluid Engineering, Cranfield,
Bedford, England, Volume 2, Paper No., Pages,
(October, 1978)

Some of the Symposium participants visiting the
vertical axis windmill at Fokker-VFW.

CONTENTS

Additional Papers:

2nd International Symposium on Wind Energy Systems
October 3rd–6th, 1978
The Royal Tropical Institute, Amsterdam

LIST OF PAPERS PRESENTED

LIST OF SESSION CHAIRMEN

Session A	H.S. Stephens, BHRA Fluid Engineering, U.K.
Session B	(Papers B1 – B4 and B8) J. Pelser, Netherlands Energy Research Foundation, E.C.N.
	(Papers B5 – B7) O. Ljungstrom, Aeronautical Research Institute of Sweden
Session C	(Papers C1 and C2) R.J. Templin, National Research Council, Canada
	(Papers C3 – C6) H. Selzer, ERNO Raumfahrttechnik GmbH, Federal Republic of Germany
Session D	S. Hugosson, National Swedish Board for Energy Source Development
Session E	L. Divone, Department of Energy, U.S.A.
Session F	P.B.S. Lissaman, AeroVironment Inc., U.S.A.
Session G	Th. Ykema, KEMA, Netherlands
Session H	P.J. Musgrove, Reading University, U.K.

OPENING SPEECH

A.A.T. van Rhijn

Deputy General Director for Energy

Ministry of Economic Affairs, Netherlands

Ladies and Gentlemen,

It is with much pleasure that I have accepted to say a few words opening this symposium on Wind energy Systems today. Especially because this symposium takes place in the Netherlands, which has a long tradition in the use of wind energy. Those who will attend the lecture of Thursday may see the different applications of the old Dutch windmills. Another application of the wind having a long tradition in the Netherlands, sailing as a means of transportation. In the 17th century these uses of the wind were the basis of the Dutch welfare. This shows the vital role of energy for a nation's economy.

It is today nearly 5 years ago that a special kind of storm rose on the world: the start of the energy crisis. To be precise at 16 October 1973 most of the oil exporting countries decided to nearly double their prices of oil. This international storm was followed by a kind of national political storm; the Netherlands as some other countries, were faced with an oil embargo. At 1 January 1974 another hard wind blew in the face of the world: the price of oil was doubled again. I recall this because if this had not happened you probably would not have been here and certainly not in so great numbers; after these storms wind energy became a topic.

Since 1973 a great number of fundamental changes in the energy policy of most countries took place and many new activities have been started especially in the area of alternative energy sources. In this 5 year period much has been established already; yet there remains still a lot to be done.

In the discussions on the development of the long term energy situation the contribution that can be expected from new energy sources plays an important role. Although it will take in any case a number of years before a noticeable contribution to the energy supply can be realised, at this moment the contribution is still uncertain. At this moment much of the R and D effort is devoted at resolving this uncertainty. This is also the goal of the Dutch National energy research programme on wind energy. This programme was approved by the Government in February 1976. It is financed by the Ministry of Economic Affairs and managed by the Netherlands Foundation for Energy Research (ECN). The objective is to study the technical and economic feasibility of large scale utilisation of wind energy in the Netherlands by generation of electricity to fed into the utility grid. The programme should be accomplished during a 5 year period and the budget of this programme was f 15 million. I shall not go into details of this programme because this will be done by Mr. Piepers of the ECN in the next lecture and a number of other speakers later on at this Conference. The main reasons for the Government support to this research programme are not only the possibility that wind energy may be able to contribute a few percent to our energy supply but also the interest of the Dutch industry for this subject. In the framework of this programme the participating industries and research organisations have completed a design for an experimental 25 meter horizontal axis wind turbine. The total costs of it are estimated at $f6,9$ million. This experimental turbine should provide the information necessary to be able to design and build wind turbines of larger dimensions.

It is a pleasure for me to be able to announce today at the beginning of this conference, that this week the Minister of Economic Affairs has approved these plans and for the financing of this project has decided to increase the budget of the national programme with f 3,5 million in order to make the building of this experimental wind turbine possible.

The Dutch Government is very much aware that, as most other countries, the Netherlands cannot solve their energy problems alone, without international cooperation. So it is striving for a close international cooperation. In this respect the most important organisations are the International Energy Agency and of course the European Community.

In the IEA cooperation the Netherlands is the lead country in the working party for wind energy. As you all know in the autumn of 1977 these efforts have led to an Implementing Agreement in which BDR, USA, Canada, Denmark, Ireland, New Zealand and the Netherlands cooperate. This year Japan, Austria and Norway joined and presently the U.K. is considering participation. In the European Community in 1975 a four year programme was started on new energy sources. This programme up to now concentrated on conservation, solar energy, geothermal energy, hydrogen production and systems analysis. In the plans for the next four years programme, 1979-1983, the inclusion of wind energy is being considered. We are strongly in favour of such an extension and we will surely stimulate as much as we can that this plan will be realised.

Although in the recent years great changes have occurred in the R and D policy of most Governments and certainly of the Dutch Government, the importance of these changes at least in my country have not always been appreciated. Too often the feelings are that attention is still focussed far too much in the nuclear field. This is certainly not true any more. This can most clearly be seen when one considers the Dutch budget for energy R and D more in detail.

In the Netherlands until 1974 the most part of the Government sponsored R and D was indeed concentrated in the nuclear field. Less then 10% of this effort was not directly or indirectly related to nuclear energy research. At present, according to a recent survey made in cooperation with the Minister of Science Policy, the non-nuclear research in the Government energy research budget accounts to 70% on a total of f145 million. In this calculation indirectly related energy R and D, such as the true fundamental research, including nuclear fusion, and industrial R and D, such as UCN and SNR-300, were excluded, because the governmental support is not primarily aimed at solving the future energy problems of the Netherlands but rather at supporting Dutch science and industry to play its role. But even if one includes the energy related research in the calculations and also the Dutch contribution to the energy research budget of the EC, then the non-nuclear efforts amount still to 40% of the total budget in 1977 and 1978. This implies that in a period of 4 years the relative share of the non-nuclear research effort has increased with at least a factor 4. The main increases in the Netherlands have occurred in the areas of energy conservation, coal R and D and new energy sources. Comparable figures exist for many other industrialised countries according to a recent survey of the IEA, that was published a few months ago. In judging these figures it should be realised that although the absolute amount of support to renewable sources may still be small, the relative share in the budgets is presently in most countries more than the anticipated contribution of these energy sources to the energy supply around the year 2000. In the Netherlands the share of wind energy in the total budget is about 5% and the maximum contribution of wind energy is estimated to be less than 4%.

I hope that this conference will be successful and I wish you all a very productive conference and a pleasant stay in our old "Mill Country".

AFTER DINNER SPEECH

Professor V. Hutter

Institut fur Flugzeugbau, Stuttgart University

Federal Republic of Germany

Ladies and Gentlemen,

It has been a sensible decision to choose Amsterdam as the venue for this BHRA wind energy systems symposium - as the Netherlands is the cradle of the most successful types of wind energy converters ever operated on this globe so far.

When developing sophisticated systems it is always best to stand on two legs. One leg is, careful reflection on the past; the other leg, speculative dreaming into the future.

Dipping deep into the past, we learn that almost nine centuries ago - during the first crusade the European knights, apart from discovering candied fruits and decimal figures, discerned the importance of those strange sailwing windmills.

There does exist a document, dated 1105, six years after the end of the first crusade, which granted the monastry of Savigny a windmill previlegium.

Only 4 centuries later, after many improvements, at the threshold of the Baroque, the Dutch windmill reached its full beauty, economy and power.

Their almost entirely wooden shaft and spars, as well as the gears with replaceable wooden teeth guaranteed a lifetime of almost 200 years.

The originally soapstone shells of the rotor's mainshaft bearings offered excellent emergency lubrication and, open on the top, a remarkable failsafe behaviour if - in severe storms - the system once should come out of control.

The miller's house and workshop containing the mill itself, was also the tower; thus not upsetting the economy balance.

Moreover the biggest of the late Baroque Dutch Mills had rotor diameters, of 26 up to 28m bigger than most experimental wind electric units installed and operated since the twenties of our century.

Only the Smith/Wilcox/Putnam 1,25 MW unit, operational on Grandpa's Knob in Vermont, USA since 1941, with a diameter approximately 54m has been of significantly bigger diameter than the Dutch Mill.

The Peril/Sabinin/Fateew d.c. unit of Balaclava, operational in 1936, had 30m rotor diameter and 60 kW rated power.

Also our 34m rotor diameter stotten-unit with 100 kW a.c. generator in 1957/1968 surpassed the 18th century Dutch Mill dimension.

The 600 kW experimental 3-blade a.c. unit of the Electricite de France at Nogent le Roi and Johannes Juul's Gjedser plant with 24m rotor and 200 kW a.c. rated power had approximately the same swept area size as the baroque mill.

I have so far not mentioned Halladay's multiblade steel mill, presented at the World Fair 1876 in Philadelphia, though it was the first wind energy system with fully automatic control.

With less than 3m diameter, it was incomparably small than Dutch windwater-pumps. However it was very successful, being adapted to the maximum water productivity of tube wells, especially in semi-arid areas. Almost 300 000 such units are still operating, especially in Australia, South Africa, USA and South America.

Already quite early analytic approaches to wind turbine theory led to insights and improvements: Bernoully, Euler, Coulomb and Smeaton performed first calculations of output/windspeed correlations.

Smeaton operated in addition a turnaround to prove its results. He reported 1756 to "The Royal Society of London" about his tests.

Remarkable is the work of Paul La Cour in the first decades of the 20th century. He tested at Askov in Denmark wind turbine models in an Eiffel wind tunnel.

The impact of results from aeronautics research enabled Prandtl, Betz and many others to improve further the theory of modern wind turbine rotors, also with comparatively high rated speed, early in the twenties.

First field tests with several wind turbines operated late in the twenties. Kurt Bielau presented his 4-blade 10m Ø ventimotor with airbrake control and his "Drehheck" design.

Big meetings were organised on the topic "New Energies for Semi-arid Areas of Developing Countries" in New Delhi 1954 and in Rome 1961. Wind energy research contributed there a lot.

At that time I met Argand and Golding, Arthur Stodhart, Johannes Juul and Vadot of Nepric in Grenoble. Andreau's fantastic depression type design of a gearless wind generator operated at Enfield Cables and in Algeria.

E.W. Golding of the ERA called a meeting in London, where he reported about wind registration over hills. It was a time of open and cordial exchange of ideas and hopes, speculations and experiences between the wind energy enthusiasts.

Reflecting "the other leg" now.:
If somebody should ask me about my opinion concerning the future of wind energy worldwide - considering such a background - I would comment: "very optimistic".

A broader and growing interest in integral wind-solar-energy, heat/pump/heat storage systems could rise and rise - if once reliable wind energy converter units of adequate size were available in quantity on the market.

Three capacity - that means magnitude - categories will have realistic chances on a real open market.
First: Units, big enough to supply the energy water, heat, lights, small machinery and communication for farm families; that would be a unit with 100 up to 200m^2 swept area in the wind, d.c. storage and a.c. supply.
Second: Units of 1000 up to 3000m^2 in the wind for similar purposes, but for communities, especially in developing countries.
Heat can easily be stored, water also.

The remaining demand usually varies with day, week and season. Doubtless that demand can easily be adapted to the availability of energy.

Third: Bigger units for public network operation, supplying the demand of industrial factories must have disc areas as big as nations with experience in highly developed technology can produce.

The central engineering problem of such 5000 to 50 000m^2 swept area in the wind units is the realisation of highest possible rated speeds.

Noise and optical pollution of the landscape is no problem. Big rotors need 4 to 12 seconds and more for one revolution. If grouped in a territory, a movement slow and majestic with different angle positions of each rotor will appear on the landscape.

High rated speed rotors need global optimisation with respect to several parameters:

Firstly: Reynolds and Mach numbers that effects the lift to drag ratio,

Secondly: Rotor-Tower Resonance safety tuning with respect to disc-wise and/or windwise blade excitation; damping and limitation of structural stresses.

Thirdly: Evaluation of the rotor's aerodynamic characteristics - taking into account also vortex state effects, starting and developing from the rotor tip on or within the disc. Frame conditions have to be: that a rotor should be optimal, if operating on his c_p/λ_ω envelope, being close to stall as well as close to vortex state locally on the blades.

Fourthly: Lifetime strength of material, due to the considered load cases and - development of load cases with respect to unloading by centrifugal forces and by surpassing of maximum lift coefficients.

Fifthly: Global control of big units is a by far less difficult and a specifically less expensive problem than with very small units.

Sixthly: Positioning of a horizontal axis turbine into the wind direction is neither expensive nor problematic. The excellent solution of Lee, Fleury and Castelli, (1780) proved to be valuable also in our century.

Seventhly: With respect to mass minimisation it is wise to make the maximum use of any element - if possible a double use of certain elements. E.g.: Energy transport for example from rotorshaft to the ground by a loaded tower supporting system.

Since big oil well drilling and operating platforms proved to be feasible also in European seas, offshore multi-unit fields won attractivity and seem to be feasible as we heard in one session from two of the 61 speakers of that meeting.

Reflecting what we experienced here at that meeting, we have reason to be thankful to all these speakers for their ambitious and comprehensive work. Because it is most deserving to contribute not only by the work performed so far, but also by the open exchange of experiences and opinions, what promotes science and engineering potential of our technology - so friendly to all human beings.

CONSERVATION OF WIND - AND WATERMILLS IN THE NETHERLANDS

Notes from an illustrated talk by **Mr. A.J. de Koning**

Technical Adviser, De Hollandsche Molen, Netherlands

General

- The Dutch did not invent the windmill.
- This kind of grinding machinery was already known in the thirteenth century.
- Watermills date from an even earlier time.
- Some sort of grinding machinery was already described by VITRUVIUS, the well-known Roman writer and engineer.
- It is generally believed that windmills originated in the Middle-East and that they were introduced to our countries by Crusaders.
- Obviously the windmill was most extensively developed in Western-Europe. The application both to polder drainage and early industry being of special significance.
- From material still extant, and the literature, one can see that our country is the leading authority on the differing types and applications of mills.

Watermills

- In addition to about 945 windmills still in existence, there are about 75 watermills spread over the province of OVERIJSSEL, GELDERLAND, LIMBURG en NOORD-BRABANT, clearly areas where flowing water in brooks and small rivers make their working possible.
- Interesting watermills are to be found in DENEKAMP (Singraven), BORCULO, south of EINDHOVEN (Genneper molen) and WYLRé (L.) used not only for milling grain; but also for sawing timber and extracting oil.

Windmills

- The emphasis concerning the preservation and application of mills, in our country, is on windmills.
- Historical publications also confirm the great and continuous importance of windmills.
- A closed study of windmills necessitates a differentiation between two main groups:
  a. **INDUSTRIAL**
  b. **POLDER**
- The **first category** refers to **CORN-MILLS, OIL MILLS, PAPER MILLS** and SAW-MILLS perhaps better described as "all windmills other than drainage mills".
- The second has been essential to the making of **POLDERS** and to the reclaiming of marshland.
- The first windmills - industrial mills - were mentioned already in the thirteenth century (Haarlem 1274). These were **POSTMILLS**. This type remained soley as **CORN-MILLS** until the end of the fourteenth century.
- Forty (40) examples of this historically important mill still exist in our country.
- From authoritative articles it has become clear that **DRAINAGE MILLS** were known at the beginning of the fourteenth century.
- At about the same time it is probable that **POLDERS** were beginning to be made.
- Probably the first **POLDER MILLS** were of the **HOLLOW POST** or **WIP MILL** type.
- Many are still seen in South-Holland.
- There is a distinct similarity between this type and the **POST MILL**.
- The arrival of the **steam engine** ushered in the so-called **industrial revolution** age.
- In theory this implied the end of **windpower** which had persisted for more than five (5) centuries.

- In our country nevertheless the end was not so abrupt.
- This was because of our favourable wind conditions, and also the high level of perfection to which the instrument had been brought.
- Successful competition with the **"FIREMACHINE"** could still be maintained.

Improvements of the windmill

- After the first **WORLD WAR** complete destruction of the remaining mills was feared.
- This led to the formation of the **MILL SOCIETY** ("De Hollandsche Molen") on the 15th May 1923.
- One of the first activities was the organising of a **competition to improve the mill** and at the same time to keep its beauty.
- This resulted in the development of a number of **SAIL SYSTEMS** identified with the names of:
  VON BAUMHAUER, HAVINGA, DEKKER, VAN BUSSEL, TEN HAVE, VAN RIET and FAUëL.
- Today we should realise that, at a time when the idea of subsidies was entirely unknown, thanks to the work of the above named pioneers in the 1920's, a large number of mills have been preserved.

Mill preservation to-day

- The position is completely different today.
- Compared with the cost of **labour,** that of **energy** is low.
- This increased cost, in such a labour intensive business as mill-working, has made the position more and more difficult.
- An additional factor is the change over of the role of miller from craftsman to a tradesman in cattlefoods and similar products.
- Concerning **POLDER MILLS** modern informed agricultural technnical opinion holds that the use of intermittent windpower for water storage in **POLDERS** is unsatisfactory and land was becoming more at risk.
- This **need** for **STORAGE** is decreasing rapidly.
- Today optimum control by machinery is more readily obtained.
- Out of a total number of about 1000 mills existing in Holland, 25% still work from time to time.
- The function of the implement is changing to that of a **MONUMENT.**
- "DE HOLLANDSCHE MOLEN" is in the happy position of being able to state that nearly all mills are now listed as **ANCIENT MONUMENTS.**
- The chance of their persisting is therefore excellent.
- Of even greater importance is the threat of the urbanisation of our country.
- Many a mill originally sited on the edge of a village is becoming surrounded by building.
- In future planning, an understanding appreciation of these matters must exist.
- To protect the rural setting of the **POLDER MILL** it is necessary to monitor suburban development.
- Eternal vigilance is needed to preserve the so-called **cultural landscape** in which the windmill has played such a large part.
- A happy development can be seen today in the cooperation between the **STATE WOODLANDS** and the mill society.
- In the apportionment of land the latter has been given a voice in the new order of things.
- This facility is of special importance when dealing with **"series drainage" or "pumping in stages".**
- Sometimes **works of art** which have existed for centuries dominate the landscape
- An area containing one or more mills has a recognizeable and familiar charcter which must not be spoiled.
- From the start there were no legal regulations for the protection of mills.
- This was in the hands of the Provincial Councils.

- From 1945 onwards, at the Provincial level, the need of preservation became recognised policy; but until 1961 there were no legal powers to enforce it.
- This interest on the part of the Provincial Councils made it possible to enlist the help of **Planning Departments** in coordinating individual aims and objects.
- Preliminary plans play an important part in such coordination.

- In conclusion the foundation of the **SOCIETY OF VOLUNTARY MILLERS** must be mentioned
- There are already eight hundred members, many of them youngsters.
- The aim of the society, by means of practical and theoretical courses, is to train members to become amateur millers.
- Successful candidates get a certificate of competence to take full responsibility for the running of a mill.
- Many mills owned by Councils or mill societies will in future be working, thanks to these volunteers.

- In all this "De Hollandsche Molen" plays an all-important part.
- The Society possesses fifteen (15) mills, in outstandingly good condition, spread over the whole of the Netherlands.
- It also has the means to restore mills. Many, belonging to various owners, are given in working order.
- In many cases the society advises on the restoration or is the link between the interested parties
- In short "De Hollandsche Molen" does everything within its power to keep up the **present numbers** of mills, and if possible to increase the **"count"**.

Fig. 1   The Uitwijk mill at Almkerk

Fig. 2   Nistelrode

Fig. 3   The Nieuwland mill,
Hook of Holland

Fig. 4   The Mallum mill, Mallum

WIND ENERGY SYSTEMS

October 3rd - 6th, 1978

OFF-SHORE BASED WIND TURBINE SYSTEMS (OS-WTS) FOR SWEDEN
- A SYSTEMS CONCEPT STUDY

R. Hardell,

SIKOB AB, Engineering Consultants, Sweden

and

O. Ljungström,

FFA, The Aeronautical Research Institute of Sweden, Sweden

Summary

The Swedish wind energy prospecting program, initially concentrating on land based systems, now also includes studies of potential off-shore sitings, where higher specific energy output, per available group station area is obtained. The cost of wind energy is very sensitive to median wind velosity at selected sites. Land areas with high winds and corresponding low cost wind energy are scarce in Sweden. Off-shore (OS) sites may prove to be more cost-effective and will reduce the land requirement for a given national wind energy production, or may even be needed for sufficient penetration of wind-electric energy into Sweden's network of the future.

Off-shore siting surveys along Sweden's coasts, utilizing depths less than 30 m, show availability of 5000 km^2 areas, with likely potential for 3300 km^2 in good wind areas suitable for WTS-installation. In typical cases, when moving a WTS site from a near shore or good coastal land area - to off shore location, 10-20 km, the median wind increases from 8, 5-9, 5 m/s (H = 100 m) and the specific energy output (per turbine disc area) increases by 40%.

Design and installation of OS-WTS gives the freedom of using much larger unit sizes than on land, assembled complete and towed to sites, with options of relocation. This should enhance energy economy, system flexibility, and output per available group station area. Examples of design concepts, of 7-14 MW unit capacity, both HA- and VA- (Horizontal viz. Vertical Axis Turbines) are given, with associated cost estimated, with outlooks to much bigger next generation designs, 20-30 MW units.

Current full scale prototype development in Sweden includes the option of OS-HA - concepts up to 4 MW size. Design criteria for such, requiring addition of load cases in the wave and sea ice environment, etc, vs land based units are discussed. OS means added costs of installation, etc, to be compensated by the higher energy production. At this early stage of study, a break even in this respect between land and sea based systems can be visualised.

Large WTS units installed on good OS-sites will yield high specific energy out-put. An example is given, for planning of a selected group station using 6, 5 MW size units on 100 km^2 off Öland, producing 550 MW, 1, 5 TWh (5, 5 MW/km^2, 1, 5 TWh/100 km^2) and up to 900 MW when using 15 MW size units (2, 5 TWh/100 km^2). In assessing the overall OS- wind energy availability in southern Sweden (using 7 MW size units) a proposed model is presented, using potential areas of 1300 km^2 near shore, producing 15 TWh/year, and 1000 km^2 distant off shore, producing 14 TWh/year additional wind electric energy.

Environmental, social and industrial aspects and impacts of OS-WTS are discussed, in comparison with land based systems. Some important environmental impacts of the latter are eliminated (e.g. land use, near site noise) or greatly alleviated (TV- disturbance, science interference - at least far off - shore). Some disadvantages with OS, affecting economy and environmental - social acceptance are also brought up, but the general conclusion is that the advantages vs land based WTS are dominant and that further OS-systems design development and planning for Sweden is strongly encouraged.

Nomenclature

A_d	Turbine disc area, m^2
AC	Alternative configurations (of WTS)
C_E	Cost of energy Sw. Cr./kWh
d	Spacing between wind turbines in array, m , km or distance from shore
D	Wind Turbine or tower diameter, m
FFA	The Aeronautical Research Institute of Sweden
GWh	Gigawatt-hours
H	Height above ground, m
HA	Horizontal Axis (turbines)
KTH	Royal Institute of Technology, Stockholm
LB	Land Based
mill	1/1000 of one US $
MW	Megawatts, $10^6 W$
N	Newton
OS	Off Shore
O&M	Operation and Maintenance
P_{Hi}	Horizontal ice load, MN
P_r	Rated power, kW , MW
P_{rgs}	Rated power for a group station
S_{gs}	Surface area, land or sea, m^2, km^2 index gs = group station
TA	Technology Assessment
TWh	Tera-watt-hours $(10^{12} Wh)$
$V_{m,100}$	Wind velocity m/s , index m, 100 = median velocity (annual) at H = 100 m
VA	Vertical Axis (turbines)
WEA	Wind Energy Area, km^2 (of certain range of V_m)
WECS	Wind Energy Conversion System
WTS	Wind Turbine System
WEPA	Wind Energy Producing Area, km^2
NE	National Swedish Board for Energy Source Development

1. Introduction

In Sept. 1977, the National Swedish Board for Energy Source Development, NE, published a comprehensive report on the results of the first three year phase of the Governement's Wind Energy program 1975 - 77, (Ref. 1). This report concludes that the combined potential of Sweden's wind energy resources and economic feasibility of large scale, MW size horizontal axis (HA) WTS justifies a continued and stepped up 3 year program (105 Mill. Sw. Cr.), aiming primarily at full scale prototype development (HA-types), as a basis for later decision of development and installation of a national WTS - up to several thousand MW electric power capacity, with land based HA-sytems as primary goal.

The current program includes R&D on future systems, such as feasibility studies of alternative configuration (AC-) WTS-technology, vertical axis (VA) types, off-shore based systems, etc. The AC-VA-studies at FFA are oriented to 5 - 7 MW size, land and off shore units. The first phase of an off shore systems, feasibility study involving design concepts, economy, environmental and social aspects, at FFA is due for completion in June 1978. In parallel, and in collaboration, studies of OS-WTS concepts suitable for prefabrication at shipyards are being performed by SIKOB, Kockums and KTH (Ref. 3, 4).

Preliminary evaluation of these studies has resulted in growing interest for OS-WTS as a viable alternative to land based WTS, and the currently issued NE-specification for design and installation of Sweden's full scale WTS prototypes (HA-type, diam. 70 - 90 m, rated power 2 - 4 MW) includes a section on OS-WTS, as an optional design (Ref. 6).

VA-design concepts are not accepted for the prototype program, but might be considered in later development phases, depending on findings in current AC-programs, as well as international progress in VA-technology.

Systems analysis of MW size HA-project families (Ref. 1) with cost-of-energy parametric analysis has revealed that the cost in c/kWh of electric energy produced is extremely sensitive to site wind velocity duration characteristics and markedly influenced by WTS unit size, (size range 1 - 5 MW), for land based systems. This is illustrated by the trend for C_E, cents/kWh vs site median wind velocity, $V_{m,100}$, Fig. 1, with close similarity to U. S. ERDA systems analysis results. For approx. 4 MW size, 2-bladed HA-turbine (diam. 100 m) on conventional tubular concrete tower (H = 100 m) having disc power loading P_r/A_d = 500 W/m^2, the energy cost dependence on $V_{m,100}$ (m/s) can be expressed as

$$C_E = 3 \cdot (V_{m,100}/9)^{-1,70} \; ; \quad c/kWh \qquad (1)$$

$$\text{for } V_{m,100} = 6 - 10 \text{ m/s} \qquad (1976 \text{ cost level})$$

This means for example that when V_m increases by 12,5 % from 8 - 9 m/s , the energy cost decreases by 18 % , from 3,65 to 3 c/kWh.

Cost effective wind energy production will require large WTS of up to 5 MW unit power, installed within the rather limited land areas in Sweden having high values of $V_{m,100}$, above 7,5 m/s , preferably above 8,5 m/s. Off-shore locations in southern Sweden gives improvements in V_m of the order of 1 - 1,5 m/s , vs land sites, resulting in considerably higher WTS specific energy output (typically 35 - 40 % increase). Design modifications from land versions to OS-WTS-installations will result in higher installation costs, to be compensated by the higher energy yield. There are many other aspects to be considered when going from LB - to OS-WTS, which will be discussed in the following.

OS-WTS-studies are also being performed in the U. S. A. (Ref. 5) and elsewhere (e. g. Netherlands). The associated ocean environment is however much more severe than the coastal waters of Sweden, providing thousands of square km of shallow sheltered waters (max. depth 25 - 30 m), with modest seaway conditions, suitable for installation of large farms of bottom-placed WTS, up to a total power capacity of the same magnitude as the planned LB-system (e.g. 5000 MW , 15 TWh).

2. Wind Energy Resources in Sweden's Coastal Areas

The Swedish wind energy prospecting program has resulted in fairly detailed isovent mapping of primarily near-coastal areas (Ref. 1) for heights over ground 10, 50, 100 and 150 m. Fig. 2 shows the 100 m median wind ($V_{m,100}$ m/s) isovent map (5 - 9 m/s), with shadings of land areas within the highest wind velocity limits. This clearly demonstrates that the best wind energy producing land areas are rather limited and are to be found in southern Sweden's coastal areas and on the islands of Öland and Gotland. In consequence, the best wind energy producing land areas are located in districts with high population density and/or which are socially - environmentally sensitive in various ways. Near coastal, off shore locations will give higher specific wind energy output. In the following, we shall examine the potential energy gain when going off shore.

Fig. 3 shows the distribution of land areas (WEA) in eight southern provinces - over a range of $V_{m,100}$ - values, 7 - 10 m/s , as well as the much smaller estimated disposable land areas for WTS - siting (WEPA - Wind Energy Producing Areas). It is found that the likely disposable land area for reasonably cost-effective wind energy production, having $V_{m,100} > 8$ m/s (8 - 9, 5 m/s) is only ca 550 km^2 , which at best, when installing WTS farms with large size units (4 - 5 MW) will yield 3 - 4 - 5 W per m^2 of group station area, total installed power 2500 MW , annual energy production ca. 7 TWh. In order to achieve the national goal of 5000 MW , 14 TWh of wind electric energy before the year 2000 , the less cost effective land areas, with $V_{m,100}$ = 7 - 8 m/s , ca. 1250 km^2 installed power ca. 3000 MW (2, 5 W/m^2) will have to be exploited. This gives a clear case for seriously considering the use of off shore sites, with potential for adding at least 5000 km^2 WEA , 1200 km^2 WEPA in shallow waters (less than 25 - 30 m depth), and up to 10 000 MW off shore wind power (Fig. 2). How much off shore energy output which will be economically feasible and practical is now being studied and will be discussed in the following.

Survey of Off-Shore Areas with Best Potential

In a preliminary survey 1976 of off shore areas with water depth less than 25 m, along the full length of Sweden's coast, from Strömstad via Ystad - Stockholm to Haparanda the above mentioned WEA = 5000 km^2 and WEPA = 1200 km^2 were identified, without regard to wind velocity. A later survey of suitable OS-WEPA in good wind areas (Ref. 8) has yielded the proposed sitings shown in Fig. 4, with total surface area 3300 m^2 (thereof 1300 near shore, 2000 distant off shore), and $V_{m,100}$ = 8 - 10 m/s . See further detailed analysis of these sites, ch. 5.

Favourable Wind-Velocity-Height-Profiles (WHP) Off-Shore

Fig. 5 shows typical median WHP for inland, near coastline and off shore sites in Sweden. It is clearly demonstrated that off shore sites give much less vertical wind shear than inland sites. Turbulence and gustiness decreases markedly off shore.

Wind Velocity and Power Output Increase with Off-Shore Distance

The number of good wind measuring stations located off shore are limited and therefore wind data for sites located 10 - 20 - 40 km off shore are uncertain. Ref. 7 , however gives some estimates for two representative OS - sitings at 20 km and 45 km from the nearest coastline (Öland S. Grund and Grundkallen respectively). These data and some analyses of $V_{m,100}$ - profiles across-coast-line from Fig. 2 , have been plotted in Fig. 6. As a preliminary guideline, showing a band for $V_{m,100}$ vs distance from the coast line, at different typical locations for average conditions in southern Sweden, at flat coastlines, the following rough trend equation is proposed,

$$V_{m,100} = 8 + \frac{d}{20} \text{ m/s} \pm 10 \%$$ (2)

where   d = distance off shore (positive seaward) ,   km

and max. $V_{m,100}$ = 10,3 m/s   (Grundkallen).

This is of course depending on fetch terrain models, as indicated in Fig. 6. Further research in this area is urgently needed.

Increase of turbine specific output (P_r/A_d) when moving a given WTS-unit from the best land/shoreline sites at $V_{m,100}$ = 8,5 m/s to typical off-shore sites at 9,5 m/s will be a factor of $(9,5/8,5)^3$ = 1,40 . This will however partly be counter-acted by increased WTS installation cost, as discussed in ch. 5.

3.   Off-Shore Based WTS Design Concepts

3.1   Introduction

The first phases of the OS-WTS studies at FFA, Kockums and SIKOB have included a wide variety of OS-WTS conceptual design configurations, all aiming at utilizing the potential advantage of very large unit sizes off shore, 5 - 10 MW and even bigger. FFA has specialized in VA-concepts (Ref. 9), other projects are primarily HA-concepts (Ref. 4), all aiming at a parametric evaluation of the energy cost-effectiveness for different designs and sizes, also comparing land based and sea based versions of some designs.

For the concepts discussed in this chapter, the following conditions are presupposed:

o   The WTS units are built and erected for testing at shipyards. Major repairs and overhauls can be performed at shipyards.

o   The units are sea transported to preselected sites.

o   Depending on water depth at the site, the unit may be mounted on the sea bottom or on a floating platform.

o   The median wind at a height of 100 m above sea level is assumed to be 8,7 - 9,5 m/s at the sites chosen.

o   The WTS installation will be arranged in multi-unit groups. Power transmission from each unit is effected via sea cables to a joint transformer and a main cable from the transformer to land.

o   Normal service and maintenance will be carried out at the site. After a major breakdown it shall be possible to detach the unit and tow it to a shipyard.

3.2   Bottom Placed Horizontal Axis (HA-) 7 MW Unit
(Kockums/SIKOB, Ref. 4)

Basically this concept is described by Fig. 7 and the following main data, as designed for $V_{m,85}$ = 8,7 m/s (at hub height 85 m)

Turbine diameter	101,5 m	(turbine upwind of
Number of blades	2	tower)
Hub height over sea level	85 m	
Rated power	7 MW	
Wind speed at rated power abt.	15 m/s	
Blade tip speed	120 m/s	
Energy production	21 GWh/year	
(Gross, no transmission losses)		
Utilization	3000 h/year	
(at 100 % availability)		

The foundation for the WTS unit may be designed as a caisson, as illustrated in Fig. 7. The concrete caisson is assumed to be constructed outside the shipyard and delivered to a dock where the erection of tower, turbine and machinery is carried out. The complete WTS unit can be tested before delivery, after erection in the dock. For the sea transportation of the unit a specially designed temporary wall with diameter 30 - 40 m is mounted to the unit. The same technique for the towing of WTS units can be applied as generally used for open-sea lighthouses.

After positioning the WTS unit over a pre-arranged bed, the caisson is lowered by filling it with water. The temporary wall is finally dismantled and removed.

The total costs for the 7 MW WTS unit can be specified as follows, assuming a production rate of 100 units per year.

Investments (1 US $ = 4,60 Sw. Cr.)	Mill. Sw. Cr. (1978 cost level 10 % above 1976)
Turbine and machinery	10
Tower and foundation	5
Erection at the shipyard, transportation, preparation of bed etc.	2
Sea cables and transformer Contributions to a 20 km cable transmitting nominally 100 MW	1,5
Not specified	1
Total investments	19,5 Mill. Sw. Cr. = 4,25 Mill. US $
Service and maintenance costs	0,5 -"- per year
Specific WTS unit cost	2800 Sw. Cr. /kW = 610 $/kW

The estimated electric power production cost, assuming 25 years life time and 10 % interest is 0,13 Sw. Cr. /kWh or 28,5 mills/kWh. Similar calculations have been performed for 5 and 10 MW sized. At turbine diameters exceeding about 100 m the costs for the blades may increase considerably. According to the preliminary study of this particular concept, a 7 MW unit size seems to be close to an economic optimum for the median wind levels reported for several sites outside the coast of Sweden.

3.3 Horizontal Axis Twin-Turbine Floating WTS, 14 MW

This concept is shown in Fig. 8, which illustrates a situation where the water depth is limited to a maximum of 30 m. At greater water depths the centre of rotation may be localized on a floating platform, anchored to the sea bottom. The same turbine design as in 3.2 is applied, which gives a total rated power of 14 MW for the twin turbine unit.

The installation on site starts with positioning of the bottom based foundation. Then the twin turbine unit with base floats is attached to a tow-boat, and towed to the site.

In principle, the design of this OS-WTS makes it self-adjusting to changes in wind directions. It may, however, be necessary to include an active positioning system to ensure an adequate yawing control, as well as protection against ice within the flotation radius.

The cost of electric energy production is estimated on the basis of the following investment costs, and assuming a production rate of 50 twin-turbine units per year.

Investments per unit	Mill. Sw. Cr. (1978 cost level)
Turbine and machinery	18,5
Tower, foundation, floats, beams	14,5
Erection, transportation, preparation of bed etc.	3,0
Sea cables and transformer (Contributions to a 20 km cable transmitting nominally 100 MW)	2,8
Not specified	2,0
Total	40,8 Mill. Sw. Cr.
Service and maintenance costs	1 -"- per year.

The corresponding electric power production cost (assuming 25 years life time and 10 % interest) is 0,135 Sw. Cr./kWh or 29,5 mills/kWh.

The above analysis of single and twin turbine OS-WTS indicates that the twin-turbine unit does not seem to result in direct economic advantages as compared to a single unit. However, on the one hand, the group station area required at sea to produce a given amount of energy is reduced when using the larger twin concept. On the other hand, the twin concept involves more technical difficulties in this case than in a conventional bottom placed single unit.

3.4 Bottom Placed Vertical Axis WTS, "Poseidon - I", 6,5 MW (FFA, Ref. 9)

Type drawing, see Fig. 9. Poseidon is a 2-bladed Darrieus-type VA-design with turbine height-to-diameter ratio H/D = 1,5 , cantilever tower (no guy wires), and fixed blades (no pitch control).

Main data, as designed for median wind velocity.

$V_{m, 100}$ = 9,4 m/s :   ($V_{m, 80}$ = 9,2 m/s)

Turbine diameter	100 m
-"- height	150 m
"Hub" height over sea level	80 m
Max. height	165 m
No. of blades/chord-m	2/2,5
Blade solidity A_{bl}/A_{disc}	0,10
Turbine rpm	13,5
Rated power	6,5 MW
Wind speed at rated power	15 m/s
Energy production (net , 95 % of gross)	17,5 GWh/year
Utilization (net , annual)	2700 h/year

(Note: lower than for the 7 MW HA-design)

The turbine-tower assembly rotates as one unit on a large bearing in the machinery housing. All primary mechanical and electrical installations are thus located low down in the rugged tower base of cylindrical shape, with easy access from the water surface level. Blades (chord 2,5 m) of (sectioned) aluminum extrusions.

Poseidon versions have alternative types of cantilever tower structures and foundation designs. Besides the steel shell type tower shown in Fig. 9 , later versions have less solid-looking steel truss towers, with four vertical tubes $\emptyset$ 1,5 - 1,8 m (Fig. 10). Early versions designed for floating on site, using three or four base pontoons (Fig. 18), have been abandoned, for reasons of high cost and dynamic problems, in favour of bottom placed concepts for shallow water sites.

The caisson type foundation for Poseidon I shown in Fig. 9 , is basically built and handled in a similar way as the foundation for the 7 MW HA-design above (Fig. 7). However the version shown has a larger concrete caisson diameter, providing enough buoyancy for towing to site without need for an extra temporary wall.

The system and energy costs are estimated as follows, assuming a production rate of 100 units per year (installation of ca. 600 MW/year).

Investment		Mill. Sw. Cr.	(1976 cost level, as in Ref. 1)
Turbine, tower and machinery		11,4	
Turbine (blades, supports) alumin.	2,6		
Upper tower, steel	4,3		
Machinery, controls	4,5		
Base tower, sea bed foundation		3,9	
Transportation, Site Preparation		0,6	
Sea cables etc. (as in 3.2)		1,4	
Total investment, installed units		17,3	Mill. Sw. Cr. (3,75 Mill. US $)
Service and maintenance costs		0,4	-"- per year
Specific WTS unit cost	2700 Sw. Cr. /kW (590 $/kW).		

With added costs for grid attachment, assuming 25 years life and 10 % interest, the estimated electric power production cost will be 0,135 Sw. Cr. /kWh , or 29,5 mills/kWh. This is comparable to the energy cost for a land version of the Poseidon design concept having the same size turbine, rated power 5 MW.

3.5 The "Poseidon II" VA-Project, 6,5 MW
(FFA, Ref. 9)

The design concept is illustrated in Fig. 10 , in an early version using the "conventional" 2-bladed Darrieus turbine, mounted on a cantilever steel truss tower. The bottom placed foundation is similar to the one used in the HA- 7 MW, above.

The turbine proportions are modified from Poseidon I , to height/diameter H/D = 140 m/108 m = 1,30 , thus reducing the tower height by 10 m.

Main data, power and energy production are the same as for Poseidon I in 3.4 .

In a later Poseidon III - version, not shown here, using a new type of VA-Troposkien (skip-rope) turbine on the same type of cantilever tower, the design turbine wind loads are reduced considerably and thereby the tower base diameter can be reduced. This makes possible the narrower foundation tower base diameter (10 m) , shown in Fig. 10 , and also a reduced base slab diameter, to 30 m as shown (compare 40 m for HA- 7 MW , ch. 3.2). This gives the added advantage of reducing wave impact and pack ice loads considerably.

3.6 Potential Advanced Very Large OS-WTS-Concepts

This first phase of the Swedish OS-WTS study includes looking at some advanced concepts which are primarily aiming at providing very large WTS unit power 10 MW - - 20 MW and above, such as will be quite feasible with the technique of dock assembly and tow-transport to sites. Such very large units should be particularly attractive when located on distant off shore shoals, or floating in deeper waters.

For example, an enlarged twin-turbine design, similar to the SIKOB 14 MW project (Fig. 8), although not projected, might be considered as a viable concept for growth to 20 - 30 MW unit capacity. A hypothetical modification of the 14 MW size unit (design $V_{m,100}$ = 8, 9 m/s) for siting at a distant off shore group station, with $V_{m,100}$ = 10 m/s , results in rated power up to 20 MW (10 MW per turbine). Furthermore by increasing the turbine diameter to 125 m (now 101, 5 m), it might be possible to achieve 50 % higher output, up to 2 x 15 = 30 MW for such a design. The group station area needed for 20 - 30 MW size off shore units (at 9, 5 m/s) of e. g. total capacity 400 MW is considerably less than the corresponding area on coastal land, using 5 MW units. (30 - 40 km^2 off shore vs 100 km^2 land area, Ref. 2).

The Poseidon III concept, basic design 7 MW (at off shore sites of quality $V_{m,100}$ = 9, 5 m/s) is a development of Poseidon II in ch. 3. 5 above, utilizing a new VA-turbine concept, called "Vald" (by O. Ljungström) which cannot be shown in its details here due to current patent work. This concept gives promise of greatly reducing tower fatigue loads and at the same time allowing the use of slender extruded blades (small chord) for very large WTS powers, with ease of manufacture and assembly and low turbine investment cost. For OS sites of quality $V_{m,100}$ = 9, 5 - 10 m/s , WTS unit sizes (with the general layout as Poseidon II , Fig. 10), having rated power up to 10 - 15 MW are visualized.

Another advanced concept, proposed by the FFA team, designated "3-D-Triton" (Ref. 11), utilizes multiple shaft cross-flow turbines (of the troposkien type, but inclined, not VA) arranged on a very wide and stable base triangle, three-pontoon system (floating version) or three concrete caissons (bottom placed). A baseline 10 MW design has total height 170 m, base triangle width 135 m, pontoon diameter 15 m (each of three). The estimated energy output for this concept, 26 GWh/year (at sites with $V_{m,100}$ = 9, 5 m/s) is very uncertain, due to lack of performance testing. A rough cost estimate indicates possible energy cost level 0, 14 Sw. Cr. /kWh (30, 5 mills/kWh). Using similar blade design principles as for Poseidon III , the "Triton" type might reach rated unit power levels of 20 - 30 MW in coming generations of distant off-shore locations (up to 10 m/s median velocity). Here again, as discussed for large twin-HA-designs, the needed water surface level will be reduced to 1/3 of land use for a given power installation.

4. Off-Shore WTS for Siting in Sheltered Shallow Waters. System Design Aspects.

4.1 Introduction

Off-shore sites along Southern Sweden's coastlines, in Kattegatt, Öresund and the Baltic Sea can be characterized as having the double advantage of sheltered sea conditions (modest sea states) and availability of large shallow seabed areas (5 - 30 m depth). This design environment is much less severe than for ocean off-shore sitings, such as in the North Sea and the Atlantic. However, during the winter seasons in the Baltic Sea and particularly further north, in the Gulf of Bothnia, severe ice loads on our WTS must be expected. Experience from construction of modern caisson type lighthouses in Sweden is very useful, both as regards wave and ice load conditions.

Floating OS-WTS units will be considered for installations where either the water is deep (over 50 m) or where the waves are very low. In Ref. 3 , a number of floating WTS concepts were analysed, as regards stability and motion dynamics both on site, in typical Baltic Sea state and winds, and during towing transport. Three-to-four pontoon designs, such as shown in Fig. 11 and 18 , were found to be inpractical, on the one hand due to requiring large pontoon base dimensions, both width and draught (considerably more than 5 - 6 m, excluding location at sites with that depth, high cost) on the other hand due to imposing wave-excited dynamic loads on the turbine-machinery in operation.

4.2   Some Aspects of Bottom-Placed OS-WTS Design, Assembly and Installation

OS-WTS in the 5 - 10 MW unit power class (current state-of-the-art), or even bigger in the next generation, designed for assembly in docks on the coast, and siting on shallow sea beds, will alleviate some difficult problems and limitations encountered in land based designs, dominated by transport access and size limitations on large prefabricated assemblies, both at first installation and when performing major overhauls.

Unit rated power above 5 - 7 MW requires HA turbine blade lengths over 50 m , tower structures and machinery elevations over 100 m high. On rail and roads, assembly dimensions over 40 - 50 m are not practical and expensive joints must be introduced in the large blades. The curved blades of large VA-Darrieus type turbines will be especially difficult to transport. Dock assembly of complete OS-WTS sea transport to sites gives the designer complete freedom, to go as big as he wants, when considering the WTS size effect on total energy-cost-effectiveness of WTS-investment and group station area utilization, etc.

Fig. 12 illustrates the shipyard construction and delivery of a caisson mounted bottom placed WTS , type VA-Poseidon I . As regards the best design for sea transport, to site and back to dock for overhaul there are a number of possible schemes under study, involving detachable pontoons/tow units, and caisson collars, etc. (Ref. 3, 10)

Design for the OS-environment will on the one hand alleviate or eliminate some of the land based WTS designer's concerns, such as turbine and machinery noise, TV-interference, but on the other hand the OS environment will require "marinization" of subsystems with corrosion-protection. New design critera involving wave and ice loads must be considered. Different off shore wind characteristics with higher turbine disc loadings than at land sites will change the turbine/tower wind loads.

4.3   Technical Specification for Design and Installation of OS-WTS in Sea Environment, Wave and Ice Loads

Within the current Swedish-NE full scale prototype development program, technical specifications for design and installation of HA-WTS , have been issued, primarily for land based systems, but also complementary specifications for sea based (OS) HA-systems (Ref. 6) . Here, we shall briefly comment on some special features in the design specifications for OS , Ref. 6 .

Site wind characteristics for prototype may be chosen same as for land based WTS , i. e. Sturup in S. Skåne ($V_{m,100}$ = 8, 1 m/s) . This will result in applying higher surface roughness (70) than at sea sites, and corresponding conservative gust loads.

Wave load cases are defined, similarly as for lighthouse design, assuming the maximum wave expected on the site once in 50 years (H_{50}) . Typically for sitings considered in Sweden, the maximum significant wave height ($H_{1/3}$) to be expected is 5 m (roughly corresponding to exceptional waves of 10 m height once in 50 years) .

Ice pressure loads can be effects of solid ice, pack ice and drift-ice. Pack ice can grow to heights well over 10 - 15 m and has actually turned over lighthouses in Northern Sweden (Fig. 13) . Spray striking the upper parts of the foundation and tower (up to about five times the maximum wave height) can give rise to heavy ice formation (Fig. 14) , different than for land based units. For platforms in open sea of up to 25 m in diameter (D) , the following exceptional values of horizontal ice pressure loads shall be applied :

	P_{Hi}		MN (Mega Newton)
West Coast and South Coast to S. Öland	0,5 D	but	$\geqq 2$
East Coast, Öland - Åland	1,0 D		$\geqq 3,5$
-"- Åland - Kvarken	1,5 D		$\geqq 5$
Gulf of Bothnia N. of -"-	2,0 D		$\geqq 6$

Example: A 20 m caisson off Öland will have to be designed for an ice pressure of 20 MN (2000 tons) . Pack ice does not normally develop when the water depth is more than 6 m . This gives a case for selecting sites at depths over 6 m , unless the units are protected from pack ice by islands or reefs.

5.   System Performance and Cost Evaluation. Group Stations and total System

5.1   Introduction

Comparisons of the relative merits of land based vs sea based WTS will involve many factors. Here, on the issues of performance (energy output) and cost effectiveness, we shall discuss power and energy output per group station area, relative installation costs for land based WTS vs OS-WTS , total available energy for proposed off shore territories, influence of WTS unit size on energy output.

5.2   Specific Installed Power (and Energy Output) per Group Station Area vs WTS Unit Size

In the area of group station efficiency, energy output per area used, as affected by WTS-wake interference, grouping pattern, unit size, research is continuing (Ref. 2) , but we do not have all the answers, which may be different for off shore than for land sites, also different for VA-turbines vs HA-turbines. Here, we shall apply data considered to be valid for flat land areas (modest turbulence) .

Off shore locations, at 5 - 10 km from land will have the benefit of being surrounded by large fetch areas of good $V_{m,100}$ - quality, but at the same time turbulence is low, (lower energy retrieval in WTS-wakes) , perhaps requiring increased spacing between units. Anyhow, the clearly positive effect of increased WTS-unit size on specific energy output can be utilized in OS-systems.

Fig. 15 shows the trend for density of WTS farm power installation (P_{rgs}/S_{gs} , MW/km^2) vs WTS-unit size, with data taken from Ref. 2 and later research in Sweden (Crafoord) . At a given size farm (or group station) in the 100 MW class, with median wind velocity $V_{m,100}$ = 8, 5 m/s , as for good coastal land sites (approx. turbulence intensity 0, 10) , the installed power density is typically 3, 5 - 11 MW/km^2 for WTS units 1 - 10 MW , and P_{rgs}/S_{gs} is proportional to the square root of WTS unit power. When limiting farms of ca. 5 - 7 MW unit sized to total capacity 100 - 200 MW per farm, or group station (GS) this quite high specific power should be allowable.

When increasing total output of each WTS-GS to 300 - 500 MW , with multiple rows of units, the losses due to wind energy deterioration down wind in multi-row patterns will be too high unless a somewhat larger unit spacing is applied (one can also use 100 MW groups dispersed some 50 D apart) .

The trend for P_{rgs}/S_{gs} , group stations off shore, in the ca. 100 - 500 MW class, as affected by WTS unit power (P_r , MW) and site median wind velocity ($V_{m,100}$, m/s) may be approximated as follows (until further evidence can be obtained from current research) :

$$P_{rgs} = 100 \text{ MW} : P_{rgs}/S_{gs} = 8 \cdot (P_r/5)^{0,5} \cdot (V_{m,100}/8,5)^3 \; ; \quad MW/km^2 \qquad (3)$$

$$P_{rgs} = 500 \text{ MW} : P_{rgs}/S_{gs} = 4 \cdot (P_r/5)^{0,5} \cdot (V_{m,100}/8,5)^3 \; ; \quad MW/km^2 \qquad (4)$$

Equation (4) which is considered to be somewhat conservative, has been applied in the group station models discussed in the following, with selected locations from areas (1) - (7) , Fig. 4 . The range of $V_{m,100}$ = 8 - 8,5 - 9,5 - 10 m/s corresponds to installed specific power P_{rgs}/S_{gs} = 3,3 - 4 - 5,5 - 6,5 MW/km^2 .

For example, in a group station area 5a1 , S. of Öland (Fig. 4, 14) 100 km^2 , $V_{m,100}$ = 9,2 - 9,5 m/s , an installation of 6,5 MW size WTS will give 550 MW power (5,5 MW/km^2) and requires 85 - 90 units, (two groups, each with unit spacing 7 D , 800 m in triangular pattern) . Min. distance from shore 1,5 km , max. 14 km , water depth 5 - 30 m .

Increasing WTS unit size to 15 MW and installing these in the same area, will result in :

a) Installation of 550 MW with ca. 38 units, specific power P_{rgs}/S_{gs} ca. 9,5 MW/km^2 , area required reduced to 60 km^2 , here available, if so wanted, at a minimum distance 6 km , off shore, one group with units spaced 1 km in triangular pattern.

b) Installation of up to 60 % more power on the 100 km^2 area, ca. 900 MW , 62 units, spaced ca. 1,4 km in triangular pattern.

Note that the specific power 9,5 MW/km^2 on this OS site is four times higher than the recommended value in Ref. 1 for 4 MW land based units, in good coastal areas. On-land installation of 550 and 900 MW respectively on the best available sites, $V_{m,100}$ = 8,5 m/s , using the largest currently projected WTS-units, 5 MW size, would require a minimum S_{gs} = 140 to 250 km^2 , and when applying the more cautious models for 4 MW size units in Ref. 1 , the land needed would be 230 - 400 km^2 (ca. 2,4 MW/km^2 , Fig. 15) .

The corresponding specific annual energy production, GWh/year can be estimated by applying a representative annual net utilization of 2800 hours (at P_r) . Thus, for example the S. Öland 550 - 900 MW OS-site produces approx. 1,5 - 2,5 TWh/year, per 100 km^2 , when installing 6,5 - 15 MW size units.

For a national goal wind electric output of 15 TWh/year, only ten such OS-group stations and a total of 900 6,5 MW size units, or 380 15 MW size units are needed.

5.3 Cost Comparison for OS-WTS vs Land Based WTS

In this early phase of OS-systems analysis, cost data can only be based on design concept analysis and are therefore extremely uncertain. More detailed baseline designs and sizing studies in coming phases will provide improved cost data. However some factors of cost trends can be descussed here when moving from land based WTS (currently size limited to 5 - 7 MW) to OS-WTS , with increased specific power and energy output (30 - 40 %) and much larger unit capacities, 10 - 15 MW in the first generation) .

The SIKOB 7 MW unit, ch. 3.2 , would produce ca. 5 MW on coastal land sites corresponding to the selected off shore site. Thus, the total annual capital cost (including extra cost for sea cables) + O&M cost may increase ca. 40 % from land to OS version for break even in energy cost (40 % more GWh/year off shore) . Such break even is about what can be stated at this stage. Note that consultation of sea cable suppliers has resulted in favourable added costs. In the 7 MW example, for 100 MW group located as much as 20 km off the coast, the extra cable cost would be only 1 Mill. Sw. Cr. , or 5 % of the total WTS installation cost. The turbine and machinery cost will increase somewhat for OS , by required "marinization", and increased power for same size turbine, but will decrease due to simplified design (fewer complex joints, easy assembly-erection) and transportation. Tower-foundation costs, with 30 - 40 m base width caissons will increase the unit cost considerably, with great sensitivity to water depth. The Swedish off shore sites provide large areas with only 10 - 15 m depth, where the extra caisson cost vs land unit is estimated at less than 20 % of the total investment.

In the case of the Poseidon I - concept, ch. 3. 4 , Ref. 9 , an attempted direct cost comparison between the 6,5 MW OS version (cables for 20 km off shore) and a land based 5 MW equal size turbine unit has indicated that the OS-version energy cost (0,135 Sw. Cr. /kWh) might be equal to or possibly 5 % less than for the land version (assuming very little land exploitation cost). If the use of land for WTS becomes critical and adds substantial cost, the swing in favour of OS-WTS will of course be obvious.

5. 4   Possible OS Sitings and Associated WTS Power/Energy Output

A survey of possible shallow water siting areas in Sweden (Ref. 8, 10) is shown in Fig. 4 , areas (1) - (7) . Data for these are tabulated in table 1 .

In summary, the following result is obtained :

Area Category	S_{gs} km^2 (WEPA)	P_r MW	E_{ye} ca. potential TWh/year
A. Near shore ($<$ 20 km dist.) 6 locations	1330	5400	15
B. Distant off shore	1070	4800	14
Total A + B	2400	10. 200	29

There will of course be differences in economy for sites under category A , having variations of site median winds $V_{m, 100}$ = 8 - 9, 5 m/s . There will also be exclusions of some areas due to restrictions of various kinds, ch. 6 (Ref. 10, 12) , however, not to such an extent as must be expected in surveys of land sitings.

5. 5   Example Group Station Area, S. E. Öland

Studies of suitable sitings in Ref. 10 include local siting models, with an example shown in Fig. 14 , 280 km² near shore, S. E. Öland (site (5a)) . The figure is rather self explanatory, showing e. g. median wind isovents 8 - 9 - (9, 5) m/s , water depths 5 - 30 m within the four selected siting areas, largely within the 4 NM territorial border line, also showing the 12 NM border line and an shipping corridor outside the sites.

This location has a potential for 1150 MW when using 6 - 7 MW size OS-WTS , total 190 units, and over 3 TWh/year energy production.

Table 1

Selected Potential Off Shore Areas and Associated WTS Power Output, Sweden.
Water Depth Limits 5 - 30 m .

Area No.	Location	Distance Off Shore km	Available WEPA Group Station area S_{gs} km^2	$V_{m,100}$ m/s (ave)	Specific power install. W/m^2 (S_{gs})	Rated power available P_r [1] MW
A : Near Shore Areas, < 20 km from shore						
1a,b	W. Coast-Halland Varberg-Halmstad	10-20	515	8,5	4	2000
2a	Skåne - S.W. Kungshamn-Höllviken	2,5-6	10	8,75	10	100
3b	Skåne - S.E. Kåseberga-Sandhammar. -Skillinge	3-12	160	9,4	5,5	900
5a	S.E. Öland	6-13	270	8,5-9,5	4,8	1300
6	Gotland-E. Slite	3-13	45	8,2-9	5	230
7a,b	N.E. Uppland	2-20	330	8-8,5	2,7	900
	Total A : Near Shore		1330		(4,1)	5430
B : Distant Off Shore, Banks, 20 - 70 km from Coast						
1c1	Kattegatt Lilla Middlegrund	20-30	100	9,5	6	600
1c2	Stora -"-	32-40	120	9,5	6	700
4	Baltic Sea, S.E. Öland - not included in Summary, too distant S. Midsjöbanken	(70-100)	(860)	10		(2900)
5b	N. Midsjöbanken E. Öland s. point	32-70	450	10	5,2	2300
7c,d	N. Uppland Västra Banken	20-25	70	8	3,3	230
	Finngrund-Ö. Banken	30-60	330	9	2,9	950
	Total B : Distant Off Shore		1070		(4,5)	4780
	Gross Total A - B :		2400 km^2		(4,2)	10.100 MW

Note 1)   Turbine diam. ca. 100-110 m , spacing 1 km . (Triangular pattern).

6. Environmental, Social and Industrial Aspects of Large Scale OS-WTS

Technology Assessment (TA) of land based units for Sweden is an important part of the current program (Ref. 1). The TA methodology and its primary factors for land based systems has been discussed by one of the authors in a previous symposium paper (Ref. 2).

When extending WTS sitings off shore and performing a comparative TA-analysis of land vs sea based WTS, some of the important impacts of land based WTS are alleviated, but new factors are added for sea based WTS. This is briefly commented upon as follows: (Land-based in densely populated areas)

Factor in T. A.	Impact for		Comment
	Land based	Off Shore	
o Land use	High - -very high	Very small[1] - None	1)Added power lines
o Site environment noise	Modest	None	
o -"- -"- TV-disturb.	Medium - High	Low - None	
o -"- -"-wind, wave, ice	-wind/icing	Added wave/ice	
o Site Aesthetics/WTS design & Landscape/scenic - - recreational areas	Important	Near Shore: - important Distant OS: - None	
o Ecology - wild life	Some	Very little	
o Bird migration, etc.	-"-	Some	
o Safety aspects, blade failures shedding of ice	Some	Very small	
o Transport interference, air	Some	Less	
o -"- -"- shipping	-	Important	
o Agriculture, fisheries	Little	Some - - important[2]	2)Espec. dist. OS sites
o Recreation activities, boating	Some	-"-	
o Military, Defence	Little	Important	
o Industrial - Steel Construct.	High	High	
o -"- - Ship Yards	Some	Very high	
o -"- - Export	Some	Very high	
o Contribution to national electric energy	Important but limited[3]	Important additional	3)Land use and cost of energy constraint

As regards landscaping - recreational scenery impacts, this may prove to be one of the primary factors in assessing the near shore OS-siting options. Demonstration models (2 - 3 dim) of group station sceneries for a number of off shore WTS-types are being applied, such as shown in figures 17 and 18.

Fig. 17 shows a section of a 2-row (triangular array) group station of 7 MW HA-units within the field of vision for a spectator on shore, with the nearest WTS placed 500 m from the shore. The six units seen here spaced 800 m apart produce 40 MW and 110 GWh/year. Assuming 15 MW units, the same picture represents 6 x 15 = 90 MW, spaced 1,5 km apart, the nearest unit placed ca. 800 m off shore.

Fig. 18 shows a similar archipelago/near-shore scenery with six 7 MW Poseidon I type units on pontoon foundations (now obsolete), 40 MW in sight, unit spacing 800 m, the nearest unit 500 m from the spectator.

Note, that in most of the proposed potential off shore sitings, it is quite possible to leave a bandwidth of 3 - 5 km off shore free of WTS , which means that the nearest installed WTS viewed from the shore will look like the distant smallest unit to the extreme left in figures 17, 18 .

7. Concluding Remarks

In conclusion, the following salient points in this paper are brought up:

o   Cost effective wind energy production - is very sensitive to site median wind velocity, the highest values in Sweden only to be found on limited land areas in densely populated districts, thus pointing to the possibility of using off shore sitings.

o   In Southern Sweden shallow water - off shore areas with high median winds (8,5 - 10 m/s) amount to at least 3300 km^2 , (1300 km^2 near shore, 2000 km^2 distant off shore). More accurate wind data for these sites are needed, however.

o   Off shore WTS tow-transported to-sites can be designed and built at shipyards, without the size limitations imposed on land based units, making both HA units (and VA-type units in the next generation) of 10 - 15 MW size possible, as illustrated by design concepts shown.

o   OS-WTS design involves special consideration of wave and sea-ice-loads, now specified for Swedish prototypes, applying experience from lighthouse design and operation in the Baltic Sea and Gulf of Bothnia.

o   Assessment of OS-sitings with the aid of coastal site surveys illustrated in the report shows, as regards output and economy

- high specific outputs per km^2 of sea area can be achieved, up to four times higher than for land based WTS, when using large units, up to 15 MW

- a total energy output from off shore of 15 TWh/year seems economically feasible in Sweden.

o   Assessment of OS-WTS as regards environmental, social and industrial factors and their impact, justifies the following statements, on advantages

- Assembly at shipyards and towing to sites gives great flexibility in site selection, relocation possible when needed.

- This also gives excellent service ability, major overhauls being done at main plants.

- Land use is virtually eliminated.

- Many environmental problems with land based WTS are alleviated or eliminated, such as near site noise, TV-disturbance, safety aspects of possible turbine failures and ice shedding.

- Export to other countries by sea transport quite favourable.

Some disadvantages must also be commented upon, such as:

- A somewhat more severe environment at sea, causing concern in design for corrosion, wave and ice loads.

- Higher costs for network attachment, sea cables more expensive than power lines on shore.

- Accessibility of WTS units for frequent inspections may be aggravated due to bad weather at sea.

Finally, it is concluded that interest for OS-WTS inSweden is growing steadily, and is judged to have a number of merits and a great potential, in the short term at least as a valuable complement to the most cost effective land based systems, in the long term possibly as a primary supplier of wind electric energy in Sweden.

8. Acknowledgements

The authors acknowledge valuable contributions to this paper from various members of Swedish wind energy research and systems design teams, in particular Messrs Curt Olsson of SIKOB, Holger Nordin of Kockums and Johan Gärdin of KTH. Sincere thanks also are extended to Mrs Gunnel Göransson and Mrs Désirée Falsjö for excellent assistance with preparation of figures and manuscript.

9. References

1. National Swedish Board for Energy Source Development: "Wind Energy in Sweden". (Vindenergi i Sverige). NE 1977:2 . Report of Results June 1977. Summary and Appendix with Contents and Figure Captions in English. (In Swedish).

2. Ljungström, O: "Large Scale Wind Energy Conversion System (WECS) Design and Installation as affected by Site Wind Energy Characteristics, Grouping Arrangement and Social Acceptance". The 1st International Symposium on Wind Energy Systems, Cambridge, September 1976. Paper A1. BHRA Fluid Engineering, Cranfield, Bedford, England.

3. Gärdin, J: "Off-Shore Based Wind Turbine Systems". (Sjöbaserade Vindturbinaggregat). Thesis for MS Degree, in Naval Architecture. KTH, Royal Institute of Technology, Stockholm, Sweden. (1977). (In Swedish).

4. Hardell, R., Nordin, H., et al: "Sea-Based Prefabricated WTS". (Prefabricerade Vindkraftverk). Unpublished Study, SIKOB, Kockums. (September 1977). (In Swedish).

5. Kilar, L.A: "Off-Shore Wind Energy Conversion Systems". - Preview of a Feasibility Study. Third Wind Energy Workshop. DOE, Washington, Sept.1977.

6. National Swedish Board for Energy Source Development, FFA - The Aeronautical Research Institute of Sweden, SIKOB AB (Engineering Consultants): "Technical Specification for Design and Installation of Wind Turbine Systems (WTS) in Sweden". Part 2 - Horizontal Axis Sea Based Large-Scale WTS Prototype. Edit. 1 (1978-02-15).

7. Kvick, T: "Off-Shore Wind Data, Sweden". (Lämplig vinddataunderlag för prospektering av vindenergi till havs). SMHI (The Swedish Meteorological and Hydrological Institute). Working Paper (1978-01-25). (In Swedish).

8. Winqvist, F. and Ljungström, O: "Preliminary Survey of OS-WTS. Sitings along Southern Sweden's Coasts" (Preliminär sammanställning av sjöbaseringsalternativ. Lokaliseringar i södra och mellersta Sveriges farvatten). Report OL-1978:11. (In Swedish).

9. Ljungström, O., et al: "Vertical Axis Wind Turbine Systems. Darrieus-type with Cantilever Tower, Project "Poseidon". - Preliminary Project Analysis of Land and Off-Shore Versions". (Vertikalaxlade vindturbinaggregat av Darrieustyp med fribärande torn, projekt "Poseidon". - Preliminär projektanalys, av land- och sjöbaserade versioner). FFA Report AU-1416, Part 3. (1977). (In Swedish).

10. Ljungström, O. et al: "Preliminary Systems Analysis of Off-Shore Wind Turbine Systems in Southern Sweden's Coastal Areas. - A Systems Concept Study and Technology Assessment , Phase 1 ". FFA Report AU-1472, Part 2 . (To be published Sept. 1978).

11. Ljungström, O. et al: "Multi-Rotor Cross Flow Wind Turbine Unit, Type 3-D, Triton, 10 MW, Off Shore Based. - Preliminary Study, Phase 1". (Flerrotors tvärströms vindturbinaggregat typ 3-D, "Triton" i 10 MW klassen, sjöbaserat. - Preliminär projektanalys, etapp 1). FFA Report AU-1416, Part 12. (1978, Confidential). (In Swedish).

12. Ministry for Civil Service Affairs (Civildepartementet): "National Planning of the Uses of Land and Water". (Hushållning med Mark och Vatten). Gov't. Publication SOU 1971:75. (In Swedish).

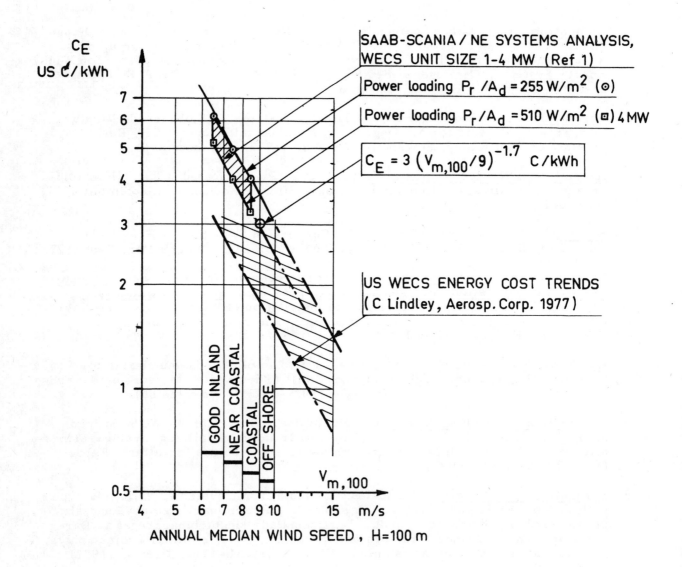

Fig. 1 Wind Electric Energy Cost, C_E ¢/kWh (1976 cost level) vs Median Wind Velocity, $V_{m, 100}$ at 100 m height. Land Based WTS (1 ¢ = 4,35 öre).

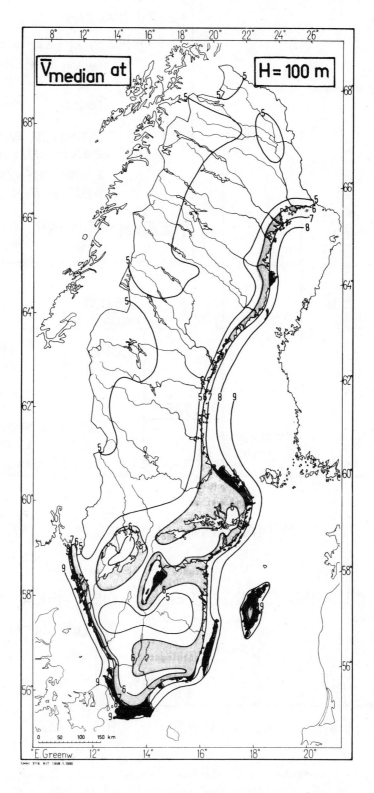

Fig. 2   Annual Median Winds at 100 m Height Above the Terrain.

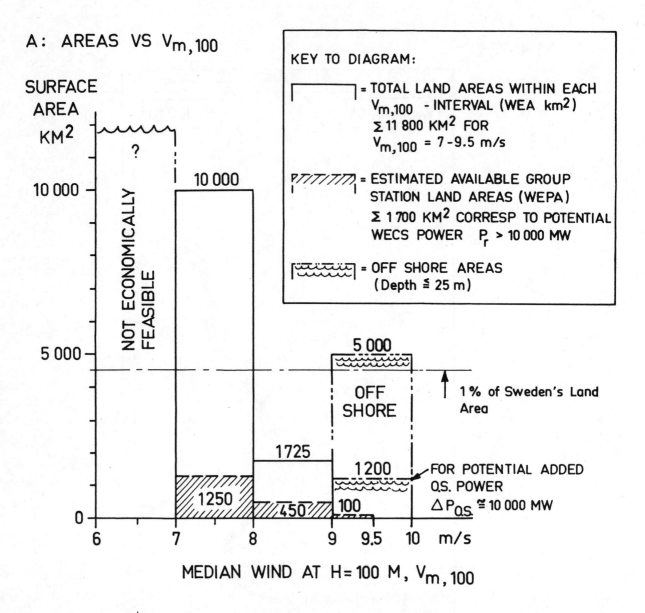

A: AREAS VS $V_{m,100}$

SURFACE AREA KM²

KEY TO DIAGRAM:

☐ = TOTAL LAND AREAS WITHIN EACH $V_{m,100}$ - INTERVAL (WEA km²) Σ 11 800 KM² FOR $V_{m,100}$ = 7-9.5 m/s

▨ = ESTIMATED AVAILABLE GROUP STATION LAND AREAS (WEPA) Σ 1 700 KM² CORRESP TO POTENTIAL WECS POWER P_r > 10 000 MW

▨ = OFF SHORE AREAS (Depth ≦ 25 m)

NOT ECONOMICALLY FEASIBLE

10 000

10 000

5 000

5 000

OFF SHORE

1% of Sweden's Land Area

1725

1200

FOR POTENTIAL ADDED O.S. POWER $\triangle P_{O.S.} \cong$ 10 000 MW

1250

450

100

0

6   7   8   9 9.5 10   m/s

MEDIAN WIND AT H=100 M, $V_{m,100}$

B: RELATIVE WIND ENERGY PER M² DISC AREA, E_A/A_d (SPECIFIC)

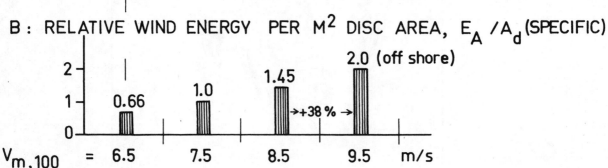

2.0 (off shore)

1.45

1.0

0.66

→+38%→

$V_{m,100}$ = 6.5   7.5   8.5   9.5   m/s

Fig. 3   Distribution of Available Land Areas over a Range of Median Wind Capacities and Estimated Associated WEPA, Wind Energy Producing Areas, for Eight Provinces in Southern Sweden. (Skåne to Uppland).

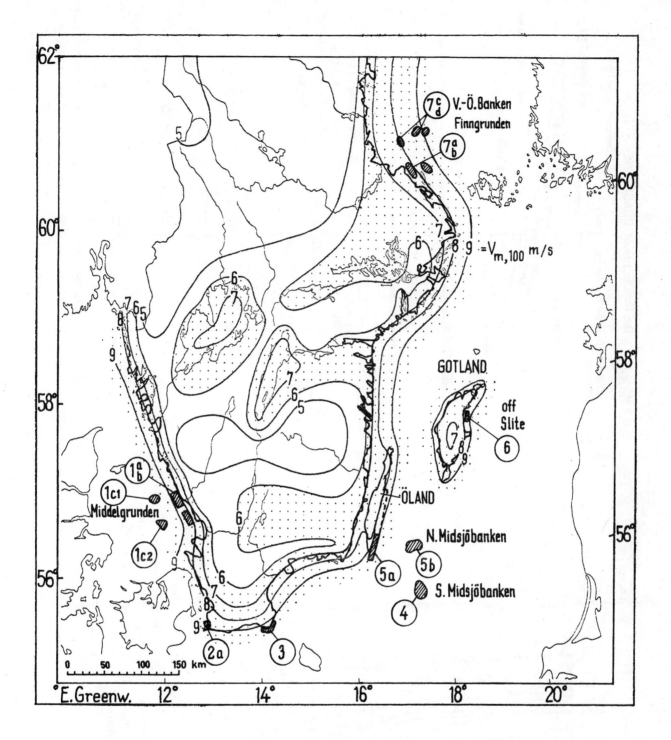

Fig. 4   Map of Some Potential Off-Shore Shallow Water WTS Siting Areas Along
Sweden's Coasts.   Total Surface Areas ① – ⑦ ca 3300 km² (depth < 30 m).

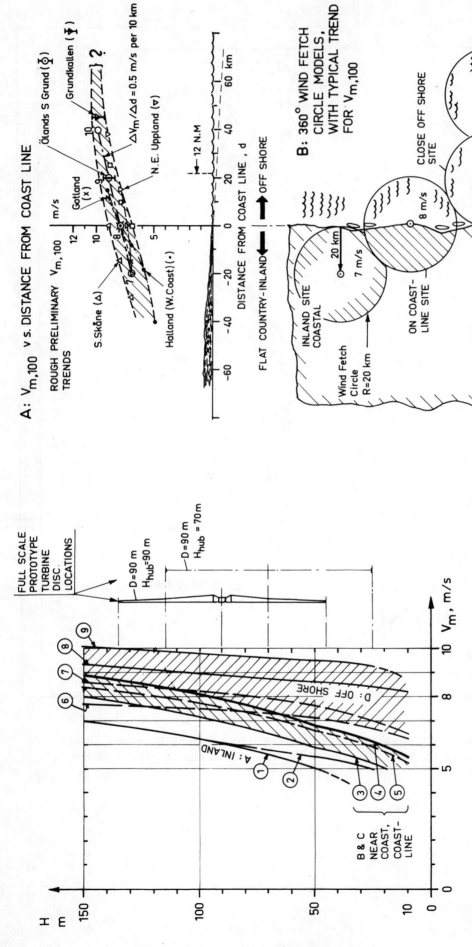

Fig. 5 Typical Median Wind Velocity Height Profiles (VHP) for A: Inland-
B: Near Coastal - C: Coastline - D: Offshore Sites in Sweden. A: Inland:
(1) Uppsala, (2) Satenäs, B: Near Coast: (3) Ängelholm, (4) Sturup,
C: Coastline: (5) Visby, D: Offshore: (6) Eggegrund, (7) Harstena,
(8) Ölands S. Grund, (9) Grundkallen.

Fig. 6 Across-Coast-Line Median Wind Velocity Trends for Southern
Sweden. $V_{m,100}$ = Annual Median Wind Velocity at H = 100 m.

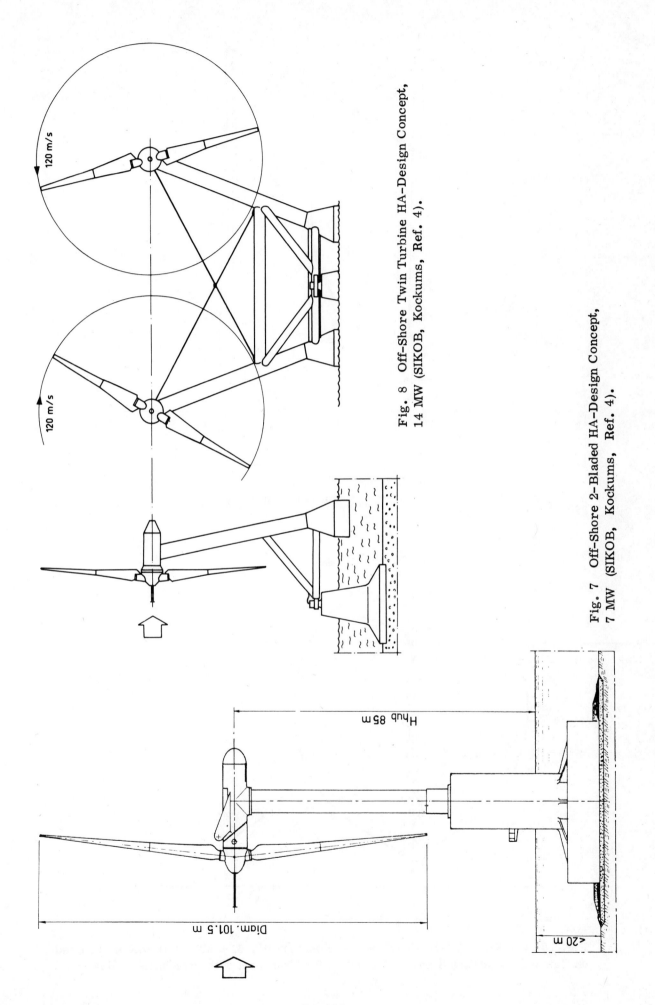

120 m/s

120 m/s

Fig. 8   Off-Shore Twin Turbine HA–Design Concept, 14 MW (SIKOB, Kockums, Ref. 4).

Fig. 7   Off-Shore 2–Bladed HA–Design Concept, 7 MW (SIKOB, Kockums, Ref. 4).

H_hub 85 m

Diam. 101.5 m

<20 m

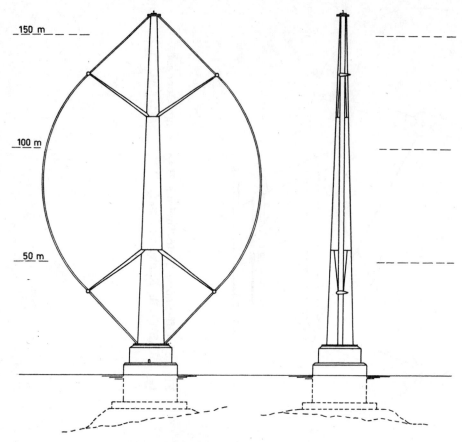

Fig. 9   Off-Shore VA-Design Concept
"Poseidon I", 6,5 MW (FFA, Ref. 9).

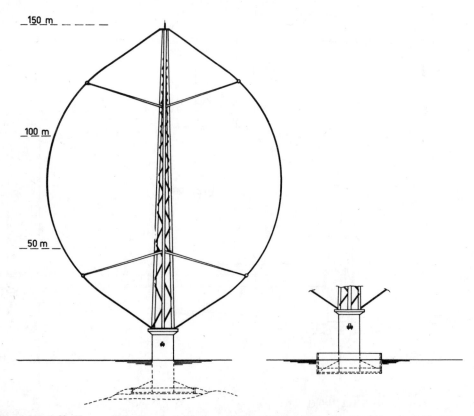

Fig. 10   Off-Shore VA-Design Concept "Poseidon II", 6, 5 MW.   a) Bottom mounted
on Caisson Foundation Slab.   b) Floating for Transport with Detachable Collar.

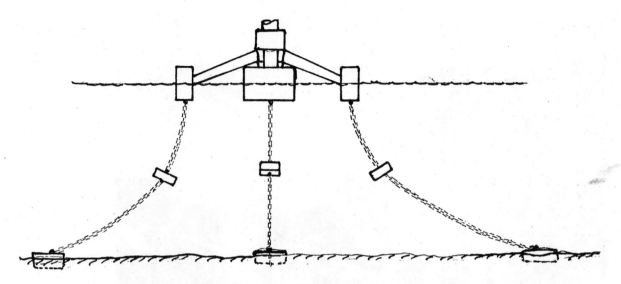

Fig. 11 Early Design of Floating WTS-Foundation with Four Pontoons and Ballasted Chain-Anchorage. Pontoon Base Width 60 - 80 m for WTS Unit Size 5 - 7 MW.

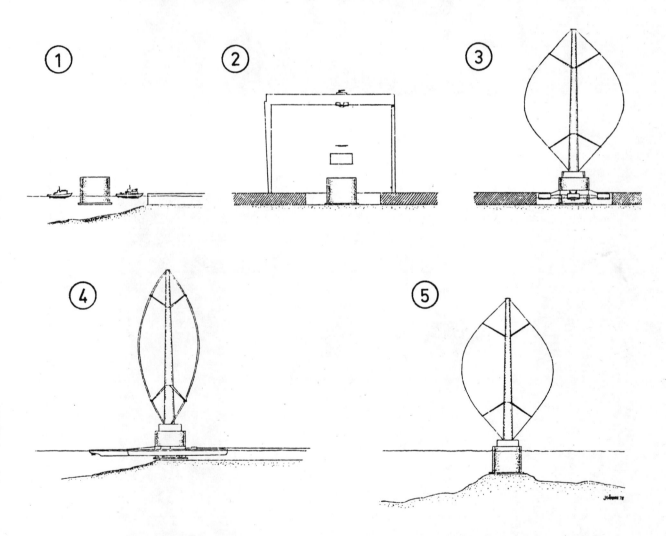

Fig. 12 Sequence of Assembly and Installation of OS-WTS, Type "Poseidon". (1)-(2) Foundation Caisson Towed to Dock. (3) Assembly of WTS on Caisson in Dock, Attachment of Float Stabilizing Pontoons. (4) WTS with Pontoons Leaving Dock, Towing to Site. (5) WTS Installed on Site, Bottom Placed, Ballasted.

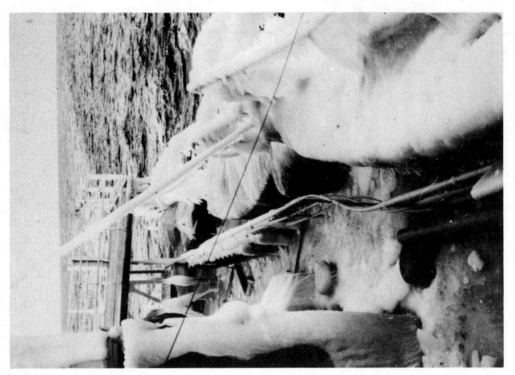

Fig. 14   Example of Icing – over of Ships at Sea.

Fig. 13   Nygrån Lighthouse off Piteå (Gulf of Bothnia) struck by severe Pack Ice (1976).
In 1970 the Tower was actually turned over by similar, even more severe Pack Ice Action.

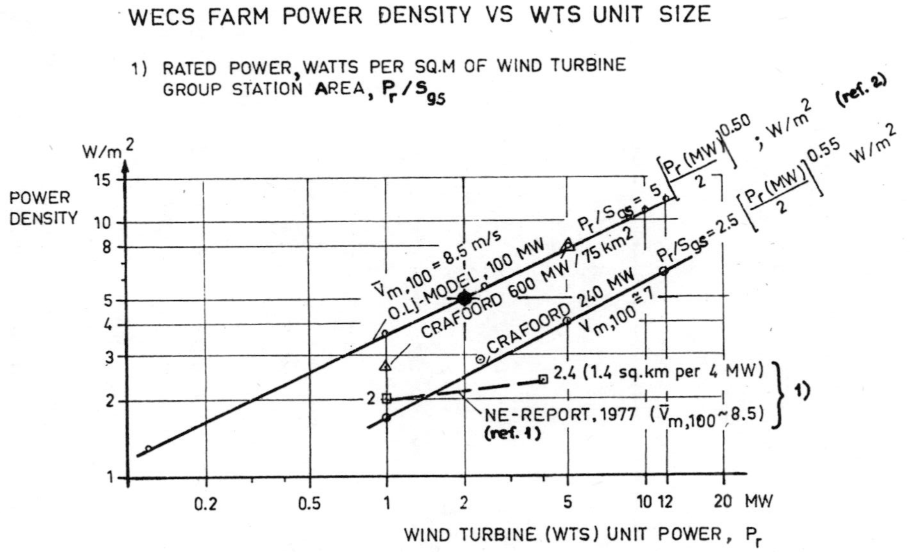

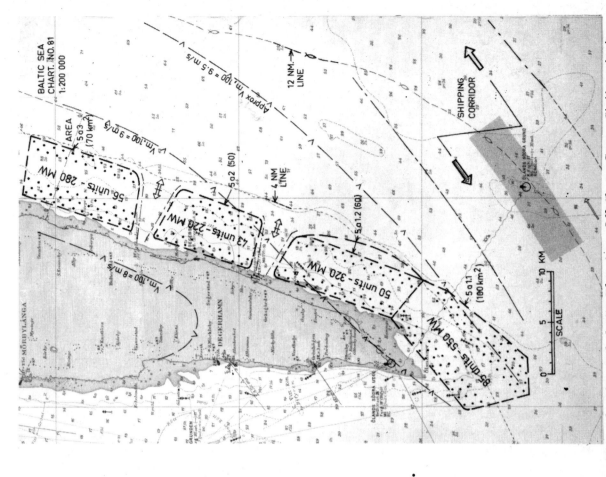

WECS FARM POWER DENSITY VS WTS UNIT SIZE

1) RATED POWER, WATTS PER SQ.M OF WIND TURBINE GROUP STATION AREA, P_r/S_{gs}

Fig. 15   WECS Farm Power Density vs WTS Unit Size – Specific Installed (Rated) Power, Watts per sq.m. of Wind Turbine Group Station Area, P_{rgs}/S_{gs} (Compiled from Data in Refs. 1 and 2).

Note 1) From Ref. 1, too conservative, without regard to actual trend for growth of power density with increasing unit size (Ref. 2).

Fig. 16   Example of Detailed Site Planning, Off-Shore Group Stations in Area ⑤ₐ , SE of Öland.   1150 MW, 190 Units, Net 6 MW Each.

Fig. 17 Off-Shore Group Station Scenery-Aesthetics, Example I : Six HA-Units, type "Windel", 7 MW Each, with Concrete Caisson Foundations on Sea Bed, Close-To-Shore Spacing 800 m. - Transport Pontoon and Tug Unit in Foreground.

Total Power Installation	40 MW	Turbine Diameter	100 m
Annual Energy Production	110 GWh	Max Height	140 m
$(V_{m,100} = 9 \text{ m/s})$		Group Station Area	$3,5 \text{ km}^2$
		(Triangular Spacing 800 m)	

Fig. 18 Off-Shore Group Station Scenery Aesthetics, Example II : Six VA-Units, Type "Poseidon I". 7 MW Each, Float-Pontoon Based, Spacing 800 m, in Outer Archipelago.

Total Power Installation	40 MW	Turbine Diameter	100 m
Annual Energy Production	110 GWh	Max Height	165 m
$(V_{m,100} = 9 \text{ m/s})$		Group Station Area	$3,5 \text{ km}^2$
		(Triangular Spacing 800 m)	

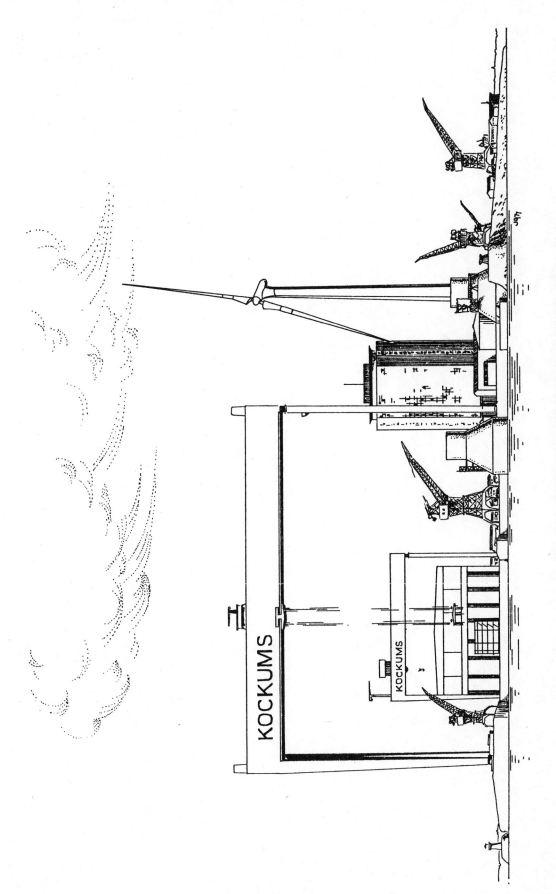

Fig. 19 Scenery of Future Off Shore Wind Turbine OS-WTS Assembly
at Kockums Shipyard. (J. Gärdin, Ref. 3 , 10).

THE EFFECT OF CONTROL MODES
ON ROTOR LOADS

E.A. Rothman,

United Technologies Corporation, U.S.A.

Summary

A limited study of the interdependence of wind turbine control modes in severe turbulence and resulting component loading has been conducted. This study was performed with a sophisticated aeroelastic, grounded model of the wind turbine with time variable collective pitch and rotor speed. The results show that turbine loading is significantly infulenced by the control mode. This suggests that turbine weight can be influenced by the control mode and, therefore, that the most direct and secure route to the lowest cost turbine is through a consideration of the complete system and its environment.

Held at the Royal Tropical Institute, Amsterdam, Netherlands.

Symposium organised by BHRA Fluid Engineering in conjunction with the
Netherlands Energy Research Foundation ECN

Introduction

The possibility of converting solar induced wind energy into electrical energy by the use of a turbine is becoming increasingly well established. The multinational representation at this conference bears witness to that.

It is our position that the technical feasibility of wind power conversion to electrical power has been well proven. Examples are the Smith-Putnam turbine described in Reference 1, the German wind turbine described in Reference 2, and most recently the ERDA/NASA MOD-O turbine at Sandusky, Ohio, USA discussed in Reference 3. The economic feasibility of wind power utilization is not as clear. A partial view of the complex economic picture can be gathered from References 4 and 5. One trend is clear, any reduction in initial turbine cost that can be achieved without a sacrifice in efficiency or reliability will be beneficial. The cost of a turbine can be estimated from its weight which, in turn, is related to the loads it must resist. Any reduction in loads that can be achieved without adding additional hardware can reduce the wind turbine cost. The results of this study show that proper use of the existing collective blade angle control mechanism can reduce rotor loads and provide an opportunity for turbine cost reduction.

Objective

The objective of this study was to determine the effect of selected modes of collective blade angle control on rotor loads. Utilizing the existing collective pitch control to minimize rotor loads could result in lower rotor weight and cost.

Scope

All of the studies utilized the turbine outlined in Figure 1. This turbine has a horizontal axis, with two blades located downwind of an open truss tower. The turbine is rated at 1500 KW of electrical power output at a turbine speed of 35 rpm and a wind velocity at 9.15 meters (30 ft.) height of 10 m/s (22 mph). The blades were cantilevered from the hub at a 12° downwind cone angle. The effective diameter of the turbine was 61 m (203 ft.). The hub was rigidly attached to the shaft which was supported by bearings rigidly attached to the nacelle. The nacelle yaw position was determined by a high stiffness double pinion drive. This turbine is typical of the types described in References 3 and 6.

Detailed studies of the wind are being conducted currently in the United States. Preliminary data are reported in Reference 7. A sample of the type of velocity variation that will be felt by a turbine in a moderately gusty wind is shown in Figure 2. The curve can be visualized as being composed of a high frequency random variation modulated by a lower frequency random variation. This study demonstrates that the rotor loads resulting from the low frequency variation can be attenuated with the existing collective pitch mechanism. Pitch modulation at a rate necessary to follow the high frequency wind velocity variations would require additional hardware and pitch modulation power. There is, moreover, a filtering of the high frequency wind velocity variations due to aerodynamic and inertial lags inherent in the rotor. This effect is discussed in Reference 8.

An intentionally severe wind velocity variation was constructed from the data and analysis described in Reference 7. A short rise time, 3.5 seconds, was selected to explore the value of control response rate in attenuating rotor loads. Previous studies have shown that blade buckling resistance during gusts is one of the primary rotor structural requirements. Buckling is a single occurrence phenomenon, that is, the blade cannot be continued in service after a single exposure to a buckling load. Therefore a large velocity change, three to one, was selected to cover the most extreme gust that might ever be experienced. The study gust, from 10 m/s (22 mph) to 27 M/S (60 mph) and back to 10 m/s (22 mph) in seven seconds, would be expected to occur very rarely, if ever, when compared with measured data from Reference 7. A more typical gust, one that could be expected once in thirty years in turbulent weather, would be 10 m/s (22 mph) to 20 m/s (45 mph). The study gust, while admittedly severe, is not unreasonable as a buckling load condition.

The control modes studies are summarized in Figure 3. Two on-line control modes were studied. The first control had a moderately fast response rate, 2.0 rad/sec., expressed as a crossover frequency. This control rate would be combined with a low stiffness connection between the rotor and the generator mount to avoid amplification of the first in-plane collective rotor mode. The second control mode has a control rate an order of magnitude slower, 0.2 rad/sec. This control rate is required for system stability with a high stiffness connection between the rotor and generator mount.

Three types of "off-line" control modes were studied. In all three cases it was assumed that the generator would come out of synchronization at 250% of rated torque, and that the tubbine would be feathered. Two constant rate schemes were studies, one at 8 degrees/second and the other at 16 degrees/second. The other scheme was a variable blade angle rate based on sensing and holding a constant rate of deceleration (negative torque).

Analytical Method

The analysis used in these studies is a special adaptation of a helicopter rigid rotor analysis created by United Technologies Research Center. The most complete description of this analysis in open literature is given in Reference 9. The adaptation of this analysis to horizontal axis, two bladed rotors mounted on an elastic tower was less difficult than with many rotor analyses. The high degree of elastic coupling inherent between a rigid rotor and its fuselage had obviated the usual assumptions of equivalent spring-mounted, rigid flapping blades and/or shaft fixity. The rotor had been treated as a part of a complete system composed of the rotor blades, hub, and a six degree of freedom mounting structure. The rotor speed was free to vary in response to internally calculated variations in turbine power output and externally imposed variations in load, and a prescribed variation in collective pitch with time. All of these capabilities were necessary to pursue this study. The most significant characteristics of the analysis, called F-762, are listed in Figure 4. It is appropriate to highlight several of these capabilities. First, the ability to include the non-coincidence of the aerodynamic center, chordwise center of gravity, and the elastic axis along the blade length is important since several independent studies have shown that lower cost blades can be built when strict adherence to classical quarter chord coincidence can be avoided. The stability of a dynamically deforming blade with non-coincident axes can only be assured by including the aero-elastic coupling that results from non-coincidence and a sufficient number of blade vibratory response modes. Seven modes are included in F-762; three flatwise, two edgewise and two torsion. In addition, the inflow velocity can be varied with time and blade azimuth to handle wind shear, full disc gusts, and partial disc gusts as well as other partial disc phenomenon such as tower shadow. Finally, the output data is very complete. Blade and hub moments, stresses and deflections are provided in the rotating system, as well as moments, shear loads and torques in the fixed system, all as functions of time.

F-762 was one of several horizontal axis wind turbine aeroelastic computer codes that was included in a thorough study of available analytical methods. This study was conducted by NASA Lewis Research Center and was reported on in Reference 10. Calculated results were compared with measured results for two steady state operational conditions on the 100 KW MOD-O wind turbine. The comparison included cyclic loads, peak loads and harmonic content. All of the loads calculated by any of the codes fell within a 1σ band about the nominal test data with one exception. All codes predicted edgewise cyclic shank moments lower than nominal for the single yaw, drive high tower blockage running.

Since that comparison was made, a refinement has been made to the definition to F-762 of the aerodynamic environment created by these two operating conditions. Specifically, the induced tangential flow description was improved, and a recalculation of loads has resulted in better overall agreement between prediction and measurement. A summary of these later calculations is included in Table I. All of the calculated values are within a 1 σ band of the nominal measured values presented in Reference 10 for both operating conditions. All of the calculations presented in this study were made with this latest verified method.

Results

The results of this study will be presented in two parts; first, the operation of the system with the various control modes operative under the gust condition will be described; second, the blade moments, stress and deflections will be presented.

Figure 5 shows the system response with Case A, the responsive control. The rate of blade angle change is rapid enough to keep the overtorque below 250% so that the system stays on-line. The percent error in rotor speed is less than 3%. It should be mentioned again, that this is a severe gust. With the slow response control, Case B, the 250% overtorque limit is exceeded. The system response is shown in Figure 6 combined with an 8°/second emergency collective pitch rate initiated after the generator comes off-line at 250% rated torque. The The rotor overspeed to 140% rated is considered unacceptable. The emergency collective pitch rate was increased to 16°/second for Case C, and the overspeed was reduced to 120% as shown in Figure 7. This was judged acceptable but the rate of rotor deceleration from 120% overspeed appeared very rapid around 4 - 6 seconds following the gust initiation. A constant deceleration rate, Case D, was then tried with only minor changes in system response as shown in Figure 8.

Turning now to resulting rotor loads, sample time histories of blade stress response with the responsive control are shown in Figure 9. Stresses are plotted at two positions on the blade, at 0.5 blade radius and at 0.8 blade radius. Examination of the stress distribution along the entire blade span shows these two radii to be most highly stressed. Bending about the chordline (flatwise) and in the plane of the chordline (edgewise) are shown. The flatwise bending response to the gust is obvious as well as the typical vibratory response to the velocity degradation of the tower. The edgewise response shows the typical gravity response prior to the gust. During the gust the response is dominated by a high frequency second edgewise response to the tower velocity degradation effect. Similar effects are shown in Figure 10. The in-plane root moments are dominated by gravity effects that change little with the gust. The steady out-of-plane root moment increases with wind velocity and the cyclic is dominated by the tower velocity degradation effect. The tip deflections of both blades are also shown with the proper phasing. The similarity between tip deflection and out-of-plane root bending moment is clear and as expected.

Identical data were calculated for the other three off-line operating conditions studies. In order to ease the comparison, the peak values of the pertinent blade stresses and moments occurring during the gust were determined, normalized and plotted in Figure 11. The highest magnitudes were associated with Case B for each quantity and that value is noted on the proper bar. A sampling of other pertinent load related information is presented as normalized bar graphs in Figure 12. The yawing and pitching moments are for the nacelle (fixed coordinate system) at the main shaft bearing location closest to the turbine. Here again the maximum values were associated with Case B, with the exception of yawing moment which was higher for Case C.

Conclusion

It is clear from the results of these analytical studies that significant reductions in blade loading, deflection, and nacelle loading during severe wind gust operation can be achieved through the use of a responsive control and the existing collective blade pitch mechanism. No specific recommendations on control frequency response values can be made from this limited work. The proper choice is intimately involved in the dynamic characteristics of the turbine rotor, the drive train, the generator, and its load. In addition, a detailed knowledge of the anticipated wind spectrum must exist. It is only through thorough aeroelastic modeling that the turbine can be studies analytically in its environment, and adequate information generated to support the trade-off decisions leading to a minimum cost, long life system.

References

1. P.C. Putnam, Power from the Wind, D. Van Nostrand Co., Inc., 1948

2. U. Hutter, Operating Experience Obtained with a 100 KW Wind Power Plant, NASA TT-F-15, 184, 1973

3. Ronald L. Thomas, Large Experimental Wind Turbines - Where Are we now, NASA TMX-71890, 1976

4. G. Rosen, H. Deabler and D. Hall; Economic Viability of Large Wind Generator Rotors, 10th Intersociety Energy Conversion Engineering Conference, 1975

5. B. Sodergard, Analysis of the Possible Use of Wind Power in Sweden Part 1, NASA TT-F-15441, 1974

6. R.L. Puthoff and P.J. Sirocky, Preliminary Design of a 100 KW Wind Turbine Generator, NASA TM X-71585, 1974

7. William C. Cliff, Wind Speed Change Guidelines, Letter to T. Healy from J.V. Ramsdell, 1978

8. U. Hutter, Influence of Wind Frequency on Rotational Speed Adjustments of Windmill Generators, NASA TT-F-15, 1973

9. P.J. Arcidiacono, Prediction of Rotor Instability at High Forward Speeds Steady Flight Differential Equations of Motion for a Flexible Helicopter Blade with Chordwise Mass Unbalance, USAAVLABS TR68-18A, 1969

10. D.A. Spera, Comparison of Computer Codes for Calculating Dynamic Loads in Wind Energy Workshop, 1977

Table I

Comparison of Measured and Calculated Cyclic
Moment Loads for Mod-O Data Case I & IV
(Nominal = 1.00 for all Cases)

	Moment Loads			
	Flatwise		Edgewise	
	Sta 40	Sta 370	Sta 40	Sta 370
Data Case I				
Calculated Cyclic	1.08	1.03	0.78	0.77
Calculated Peak	1.11	1.20	0.84	0.95
Data Case IV				
Calculated Cyclic	0.87	0.91	1.05	1.11
Calculated Peak	0.95	0.77	1.01	1.27

TURBINE CHARACTERISTICS

Rated Power	1500 KW
Rated Wind Speed	9.75 m/s (22 MPH)
Cut In/Cut Out Wind Speed	4.9/22.3 m/s (11/50 MPH)
Rotor Diameter	62m (203 Ft.)
Number of Blades	2 Downwind
Airfoil Type	23xxx
Hub Type	Rigid
Cone Angle	12°
Rated Rotor Speed	35 RPM
Tower Height	44m (144 Ft.)
Tower Type	Open Tabular

Fig. 1

WIND SPEED CHARACTER

Mean Wind Speed 12 m/s (26.8 MPH)
Turbulence Factor 0.2

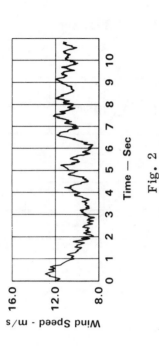

Fig. 2

CASES STUDIED

Wind Condition —

Gust - 9.75-26.8-9.75 m/s @ 9.15m, 7 Seconds, Cosine

Control Modes —

Case A - Responsive Control, 2 RAD/Sec

Case B - Slow Control, 0.2 RAD/Sec
Off-Line Control, 8°/Sec to Feather

Case C - Off-Line Control, 16°/Sec to Feather

Case D - Off-Line Control, Deceleration (N) Limiter

Fig. 3

F-762 PRINCIPAL CHARACTERISTICS

- Blade Modeled as a Coupled, Elastic Beam with Linear Twist
- Section of Gravity, Aerodynamic Center and Elastic Axis are Noncoincident with Pitch Axis, Independent and Variable with Span
- Three Flatwise Modes, Two Edgewise Modes and Two Torsion Modes
- Four Variations in Two Dimensional Airfoil Data Over Blade Span
- Variable Rotor Angular Velocity
- Multi-Blade
- Dynamic Foundation
- Inflow Field Definition Variable with Span, Azimuth and Time
- Time History Solutions of Blade, Hub and Foundation Loads

Fig. 4

CASE A

Responsive On Line Control - Kv = 2 Rad/Sec

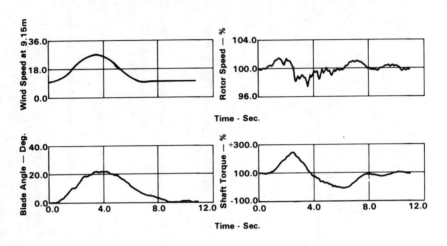

Fig. 5

CASE B

Slow On Line Control - Kv = 0.2 Rad/Sec
Disconnect from Line at 250% Elec. Torque
Slew to Feather at 8 Deg/Sec

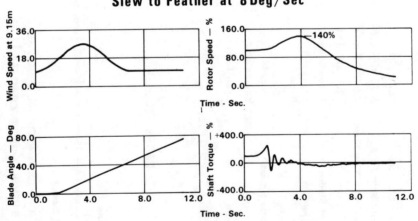

Fig. 6

CASE C

Slow On Line Control - Kv = 0.2 Rad/Sec
Disconnect from Line at 250% Elec. Torque
Slew to Feather at 16 Deg/Sec

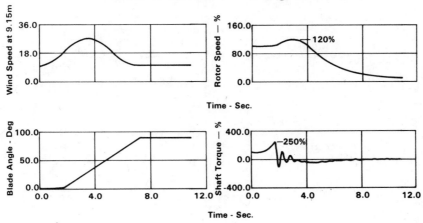

Fig. 7

CASE D

Slow On Line Control - Kv = 0.2 Rad/Sec
Disconnect from Line at 250% Elec. Torque
Closed Loop ṄR Decel Control

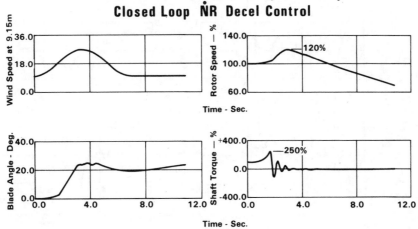

Fig. 8

BLADE STRESS
CASE A

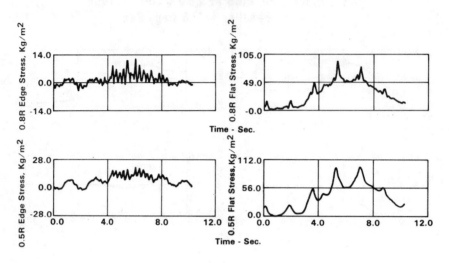

Fig. 9

ROOT MOMENTS AND TIP DEFLECTIONS
CASE A

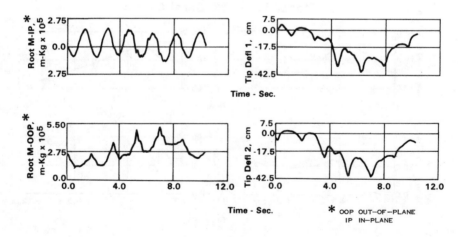

Fig. 10

LOADING SUMMARY

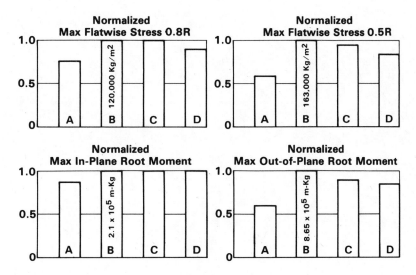

Fig. 11

LOADING SUMMARY

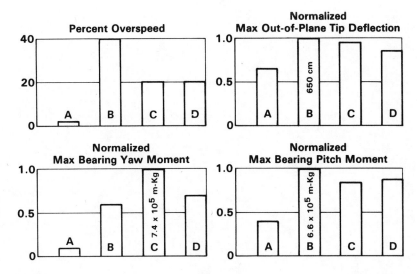

Fig. 12

WIND ENERGY SYSTEMS

October 3rd - 6th, 1978

DESIGN CONCEPT FOR A 60 m DIAMETER WIND TURBINE GENERATOR

D.F. Warne, BSc(Eng), ACGI

Electrical Research Association, U.K.

G.R. Ketley, BSc(Eng)

British Aerospace Dynamics Group, U.K.

D.H. Tyndall, CEng, MIStructE

The Cleveland Bridge & Engineering Co. Ltd., U.K.

and

R. Crowder, BSc(Eng), ACGI, CEng, MICE

Taylor Woodrow Construction Ltd., U.K.

Summary

A design for a large wind turbine generator suitable for network operation has been evolved by an industrial group comprising British Aerospace Dynamics Group, Taylor Woodrow Construction, Cleveland Bridge & Engineering Co. and ERA. The work has been undertaken in conjunction with the two Scottish Electricity Boards and with partial financial support from the Department of Energy.

The aim has been to design a machine for minimum energy cost at high mean wind speeds (10-12 m/s at 10 m height) typically found on hill sites along the north west coast of Scotland. The design philosophy has been simplicity and ruggedness; it has led to the selection of a fixed pitch, fixed speed turbine coupled through a fixed ratio transmission to an induction generator. This concept is similar to that adopted in the most successful machine yet built, the 200 kW rated machine at Gedser in Denmark.

A range of machine parameters has been examined and the 'base-line' design chosen for detailed evaluation comprises a 60 m diameter all-steel two bladed turbine mounted upwind of a 45 m high tower and rotating at 34 rpm, a fixed ratio gearbox stepping up to the generator speed of 750 rpm. Speed control is effected by the network, through inherent interaction of the turbine and generator torque speed curves. The maximum output of the base-line design is 3.7 MW and annual output, for example, at a site with a mean wind speed of 10 m/s (at 10 m) is about 10.8 GWh.

This paper describes aspects of the machine design and reviews other considerations that have been taken into account during the work.

Held at the Royal Tropical Institute, Amsterdam, Netherlands.

Symposium organised by BHRA Fluid Engineering in conjunction with the Netherlands Energy Research Foundation ECN

1. INTRODUCTION

In the UK there has been a long standing interest in large scale generation of electricity from the wind, with some notable prototype machines of modest size built in the 1950s; among the organisations involved at this time were the Electrical Research Association, De Havilland Propellers Ltd. (now British Aerospace Dynamics Group) and the North of Scotland Hydro Electric Board. It was natural that this activity should be renewed in the modern pursuit of alternate energy sources, particularly in view of the high mean wind speeds available in certain parts of the country.

In 1976 a group of industrial companies comprising British Aerospace Dynamics Group (then Hawker Siddeley Dynamics), Cleveland Bridge & Engineering, Electrical Research Association and Taylor Woodrow Construction decided to collaborate in the design and costing of a large wind turbine generator for network operation. The North of Scotland Hydro Electric Board and South of Scotland Electrcity Board agreed to assist in the project and the Department of Energy provided £75,000 for the work, which was 50% of the phase I project cost.

The requirement was a machine for use in a 'fuel-saving' mode, that is to deliver power to the network by direct connection as and when adequate wind speed is available; the economics are judged by the reduction in demand on other power stations operating at the time and the consequent saving of fuel. No consideration has been given to the design or economics of machines specifically for heating loads (which are non frequency and voltage dependent) or machines coupled to dedicated pumped storage schemes in an attempt to attribute an overall firm generating capacity. The design is therefore applicable without special considerations to most large grid systems around the world.

The main objective in the work was to progress design and application considerations of this 'fuel saving' machine to the stage of reasonable certainty concerning feasibility and economics. The following issues were specifically addressed:

- the configuration and cost of a machine optimised for minimum ouput energy cost.

- aspects of access, construction and network connection associated with practical sites.

- the number of machines and energy production likely to be economic on a national (UK) scale.

At the outset it was appreciated that full optimisation was impractical because of the necessity to undertake detailed work leading to costing on each alternative. The importance of proceeding to a single design for detailed work required careful consideration of the selection criteria and the following rationale was evolved.

First, the scheme must be simple, related as closely as possible to existing engineering practices and offering maximum potential reliability. Any complication of the turbine by moving surfaces, hinges or sophisticated materials must be avoided unless positively justified on commercial or technical grounds for the overall system.

Secondly, the machine should be designed for high mean wind speeds; optimisation against more plentiful lower mean speeds would result in an uneconomic scheme.

Thirdly, since the costs contain a fixed element associated with site preparation, access and network connection, the economies of scale apply; the turbine size should therefore be the maximum commensurate with confidence in design and construction techniques, reliability and projected cost.

Fourthly, all machine components should be designed for a minimum 20 year service life, equivalent to 100,000 operating hours in the case of the turbine generator unit.

The 'base-line' design resulting from this rationale centres around a conventional horizontal axis turbine of 60 m diameter. The turbine is two bladed, fixed pitch and fabricated of mild steel; it operates upwind of the tower at a nominal speed of 34.1 rpm and produces a peak output (electrical) of 3.7 MW at 22 m/s wind speed. The turbine drive is increased in speed to about 750 rpm with a gear transmission and the generator is of asynchronous type. The turbine generator unit is mounted on a tower head platform capable of orientation into the wind, with the turbine axis 45 m above ground level. Because of the importance of local constraints and aesthetics, both a reinforced concrete and a lattice steel tower scheme have been devised. Outlines of the alternative schemes are shown in Figs. 1 and 2.

The following sections describe the individual features of the design and discuss considerations of network connection and operation.

2. PLANT DESCRIPTION AND RATIONALE

2.1 Turbine

The turbine is a dominant feature in the conception of the plant and several fundamental choices and decisions were made in arriving at the 'base-line' design.

The conventional horizontal axis style of turbine was selected in preference to vertical axis types principally because of a lack of confidence in present performance prediction methods for the latter type, and uncertainties concerning its feasibility on a large scale. Further, no detailed case studies have yet been made in support of the notion that the saving in cost of orientation gear in a vertical axis turbine would outweigh the cost of an often more complex blade and structural support system.

An initial scan of the likely economics, past designs and other projects in hand around the world indicated that the turbine diameter should be in the range 40 m to 80 m. Power output increases as (diameter)^2, and it was considered unlikely that the cost of any major component in the turbine or the associated plant would increase more rapidly than transmitted power or the square of the scale; certain items (acquisition and preparation of the site, access, network connection and controls) would increase at a much lower rate, or not at all. In the 40 m to 80 m range, few discontinuities were anticipated in the relationship between cost and diameter, resulting in a steady decrease in effective cost of output energy with increasing size. A unit much larger than 60 m would however involve fabrication and erection of components whose size and nature were beyond existing practices in civil, mechanical and aeronautical engineering; a turbine diameter of 60 m was therefore chosen for the 'base-line' design.

The effect of number of blades, rotational speed, generator rating, fixed or variable pitch blading, and of blade chord, planform, twist, aerofoil section and angular setting were all examined for their influence on annual energy output and on other parameters which significantly affect total capital cost, such as blade, shaft and transmission loads, weight, constructional complexity and ease of maintenance. It was judged that a two bladed fixed pitch turbine was likely to provide minimum energy cost despite lower power coefficients and greater oscillatory loads associated with two blades (as opposed to three or more).

Consideration was given to four basic material forms - weldable steel, rivetted aluminium alloy, wood and glass-reinforced plastic. It was confirmed that steel is most economic, the larger blade forces caused by increased weight being more than offset by lower cost and ease of fabrication; only minor compensatory savings would be made in the tower cost through the adoption of a lighter turbine structure.

The resulting structural design is of fixed pitch non-articulated type, constructed in welded mild steel to BS 4360 Grade 50. It is assembled by bolted flange joints from five main sections (hub, two mid-blade and two outer spans), each comprising a spanwise box beam to which is attached leading and trailing edge fairings; the skin is supported at intervals by transverse box ribs. The mid-span and outer sections are twisted, tapered and coned in the downwind direction.

After careful deliberation it was decided that moving surfaces should be incorporated into the turbine for shut-down and overspeed protection; these aerodynamic braking surfaces or 'spoilers extend over the outboard 30% of the span at the blade trailing edge.

The main features of the turbine are summarized in Table 1.

2.2 Transmission and Generator

Preliminary consideration of the transmission requirements soon revealed that although the turbine output is relatively modest by power plant standards, the low rotational speed results in a high torque (1093 kNm). To minimise gravitational loadings on the nacelle, and vibrational problems in the shaft it is important to restrict the primary shaft length to a minimum. The primary stage of any speed increasing transmission must therefore be close to the turbine, and since this stage constitutes the majority of the total transmission/generator weight, it has been considered prudent for simplicity and integrity to locate the secondary transmission stage and generator also on the nacelle, discarding the possibility of ground mounting.

The major alternatives for the transmission are gears, chains, hydrostatics or direct generator drive. A direct coupled generator was investigated but soon rejected on grounds of size and cost. Hydrostatic drives have the potential for variable ratio transmission, but there is little precedent for the required duty, and efficiency is markedly lower than gear or chain drives, outweighing any benefit obtained from operating the turbine at variable speed to maintain optimum performance from constant tip-speed ratio. The preferred alternatives were a roller chain drive or a gear drive; since direct precedents existed for gear drives (used in most previous large prototype wind turbine generators, and also similar duties in cement mill drives) such a drive was selected. The transmission incorporated into the base line design is a two stage divided drive, helical spur coaxial gear transmission with a nominal ratio of 35 : 750 and a 3.9 MW input rating. The total weight of this unit is 43,300 kg.

Synchronous and induction generators were considered, discounting dc generators because of additional cost and maintenance and the need for inverters, and eliminating the various types of variable speed constant frequency generators because of lack of suitable precedents and experience. The synchronous generator offers the advantage of power factor correction (which may be important, particularly when the unit is connected to the network via a long line) but is more expensive, complex and requires more control equipment.

The cage induction generator is one of the simplest, cheapest, rugged and reliable type of electrical machine and has direct precedents well in excess of the rating required. It offers the additional benefit (since it operates with a slight speed variation as the drive torque changes) of absorbing the energy of gusts into additional kinetic energy in the turbine; the synchronous generator operates at fixed speed and gust energy may immediately be translated into additional torque, resulting in a potential scability problem.

Various generator speed/transmission ratio combinations were investigated and there was found to be little difference in overall cost in the range 750 rpm to 1500 rpm, but the cost and mass increased markedly with lower generator speeds.

The base-line design incorporates an 8 pole cage induction generator rated 3.7 MW (4.1 MVA), 3.3 kV, 3 phase, with a nominal running speed of 750 rpm for 50 Hz output.

2.3 Nacelle and Tower

Alternative designs for steel lattice and reinforced concrete towers have been prepared, but both have a nacelle supporting the turbine, gearbox and generator which is mounted on the tower and is capable of orientation into the wind. Advantage has been taken of the flexurally stiff turbine design in mounting the turbine upwind, minimising 'tower shadow' effects on the blades.

2.3.1 Reinforced concrete scheme

The main features are a reinforced concrete tower and foundation surmounted by a nacelle cantilevered out from the tower face.

The nacelle has a structural steel main frame clad in reinforced plastic and incorporates a dismountable plant pallet carrying the generator, gearbox and turbine. The plant pallet is located clear of the tower (in plan) so that the generator gearbox and turbine can be fitted to the pallet at ground level and lifted into position as a complete unit by hoisting equipment permanently built into the tower structure. Ease of initial erection and any subsequent dismantling required for major maintenance purposes well outweighs the slight penalities of additional steel in the nacelle and additional moment applied to the tower.

In the initial work, the proposal was that rotation is effected through a taper roller bearing arrangement comprising an upper and lower set of opposed inclined bogies running on tracks mounted on the outer tower wall and inclined at similar angles, with propulsion by hydraulic rams. For the next phase of work alternative options will be considered for both the bearing and propulsion systems.

The tower is a hollow reinforced cylinder 45 m high, 8 m outside diameter with a 0.35 m thick wall. The diameter and wall thickness have been chosen to ensure satisfactory stress levels under operating and overload conditions and to ensure that the natural frequency of the tower is well above the turbine blade passing frequency, a feature readily achievable using reinforced concrete. The wall of the tower shadowed by the turbine is perforated in order to minimise the possibility of vortex shedding.

The tower wall is tied to a reinforced concrete raft, the top of which is at ground level. In typical ground conditions this would be approximately 17 m diameter and 2 m deep, with a rock foundation, it is likely that a footing ring about 0.5 m deep and 1.0 m wide laid in an excavated trench and tied to the rock by a series of rock anchors would suffice.

2.3.2 Steel lattice scheme

The main features are an open lattice tower surmounted by a symmetrically located and balanced nacelle.

The nacelle has a steel beam framework which supports the turbine, gearbox and generator and includes counterweight at the downwind end to reduce the bending moment. The frame is clad in removable steel sheets.

Nacelle rotation is achieved by 12 wheel assemblies running on a circular track fixed to the tower, and propulsion is by hydraulic motors mounted in the nacelle, the drive being transmitted via pinions to a ring on the tower.

The tower is relatively slender in the region of turbine shadow to limit the eccentricity of turbine gearbox and generator weights, to minimise wind blockage and for aesthetics. Below the lowest point of turbine rotation the tower widens into four legs, giving a broad base of 18 m for stability and to limit foundation loads. Internal bracing known as inverted K is used for inherent stiffness and horizontal bracing is incorporated at several levels to increase torsional stiffness. Circular hollow section was chosen in preference to angle or universal beam because it has a higher strength to weight ratio in compression, minimises wind disturbance and has better appearance. All erection connections are made with High Strength Friction Grip Bolts.

In soft ground each tower leg is fixed to a concrete block of sufficient weight to resist uplift with a safety factor of 1.4 and having sufficient plan area to ensure safe distribution of downward load. In good rock, each leg has a small concrete footing and is tied directly to the rock by special anchor bolts.

2.4 Control and Safety

The power and speed control of the machine in its normal operating mode is inherent and requires no special control equipment.

Normal speed control of the turbine generator is achieved by inherent matching of the aerodynamic torque-speed curve of the turbine to that of the network connected induction generator. The turbine and generator torque speed curves intersect in a manner which produces a speed stabilising torque imbalance when disturbed by transients in wind speed. Although this implies slight variations in rotational speed, the turbine performance can be assumed at a constant nominal speed (34.1 rpm) and the consequent power output over the operating wind speed range is shown in Fig. 3. The cut-in wind speed is 7 m/s, the turbine generator reaches the peak output of 3.7 MW at 22 m/s, and the machine is shut down at wind speeds greater than 27 m/s.

A typical control and operation sequence would be as follows:

(a)   Start from low wind speed. The turbine may not be self starting and provision is made for using the generator as a motor to provide initial torque. When average wind speed over a limited period exceeds about 7 m/s the friction brake is released, spoilers stowed and the main circuit breakerclosed; with rotation the turbine develops torque and the unit accelerates until turbine and generator torques balance. The motor inherently becomes a generator above synchronous speed and no synchronisation is required.

(b)   Normal conditions of power delivery (wind speed between 7 and 27 m/s). The power output and the speed of the turbine generator are inherently governed by the intersecting torque-speed curves. Power output fluctuations are damped and limited by small speed changes and the considerable natural inertia of the rotating system.

(c)   Shut-down at high wind speed. When average wind speed over a limited period has exceeded 27 m/s the spoilers are deployed, the generator disconnected, and at 10% full rotational speed the friction brake is applied. Once at rest, the turbine is inched into the horizontal position and turned parallel to the wind by the orientation gear.

(d)   Operation at high wind speed. The orientation gear remains operational as for (b), but with a 90° shift in the signal.

(e)   Cut-in from high wind speed. When average wind speed over a limited period has dropped below 27 m/s, the turbine is turned to face the wind and started as in (a). (Motor assistance may not be necessary).

(f)   Shut-down at low wind speed. When mean electrical power output over a limited period is negative, the spoilers are deployed and the sequence initiated as for shut down at high wind speed, except that the turbine is not inched into the horizontal position or turned parallel to wind.

(g) Operation at low wind speed. Orientation gear remains operational as for (b), but taking direction sampling over a longer period to cope with more rapid directional fluctuations of light winds.

In the event of overspeed through loss of network connection or transmission failure, or in case of excessive vibration, the spoilers are deployed and the turbine brought to rest in the normal mode. A stand by diesel generator is incorporated to maintain essential services following loss of network connection.

2.5 Network Connection

A single machine would incorporate on site transformation to 33 kV and overhead or underground 33 kV transmission to the nearest source 33 kV substation with adequate fault capacity. The required fault MVA is determined in UK by statutory limits on voltage fluctuation; the motor starting condition has been taken as the worst case and the necessary fault MVA is then dictated in this case by the starting current, frequency of starting and the permitted voltage dip. This may be a pessimistic requirement, but in view of the present uncertainties regarding the nature of fluctuations in power output from large WTGs it will be prudent to adhere to this, although the unit may not ultimately require a motor assisted start.

Groups of machines (sited on adjacent hills) would have intermediate transmission of 11 kV between each other and to a common point of transformation to 33 kV. There would then be a trunk link to the existing network. In this case the fault capacity at the point of connection need not be in proportion to the number of machines, since some diversity in starting and in power output fluctuations will occur.

3. POTENTIAL FOR EXPLOITATION IN UK

Combining the turbine generator characteristic of Fig. 3 with velocity duration curves characteristic of the wind regimes on the west coast of Scotland, the annual energy output of the machine at different mean wind speeds has been calculated and is shown in Fig. 4.

ERA's programme of hill-site anemometry in the 1950s led to the conclusion that there are between 1,000 and 2,000 hills on the west coast of Scotland with mean wind speeds (10 m above ground) in excess of 8 m/s and some in excess of 12 m/s.

Trial budget costings of the base-line design, including not only the machine cost, but also site work, access road and network connection for selected sites have indicated that the machine is economic (when judged against current accounting criteria and present generation costs) for a modest but significant number of the higher wind speed sites available. The sensitivity of the economics to the real cost of fuel is particularly noteworthy; a postulated 10% increase in real fuel cost would more than double the number of hill sites that are potentially economic, this number representing about 5 TWh pa total generation, or about 2 mtce pa saved.

4. PROGRAMME STATUS

The UK Department of Energy has now agreed that this potential is sufficiently attractive that the justification exists to proceed immediately through a detailed design phase, from which prototype construction will be immediately possible. £341,000 support from the Department of Energy for the detail design work was announced in June 1978 and work is now in progress. Provided economic and technical promise of the base-line design is confirmed in the detailed work, construction of a prototype could start in mid 1979.

ACKNOWLEDGEMENTS

The authors are grateful for the permission of their respective Company Directors to publish this paper. The assistance of the South of Scotland Electricity Board and the North of Scotland Electricity Board and the financial support from the Department of Energy is also acknowledged.

Table 1

Leading particulars of the turbine

Diameter		60 m		
No. of blades		2		
Design rotational speed		34.1 rpm		
Max. operational·wind speed		27 m/s		
Max. shaft output		3.9 MW		
Blade characteristics	radius	10%	50%	100%
	chord (m)	3.795	2.515	0.915
	angle*	20.00^{o}	4.43^{o}	-1.25^{o}
	aerofoil section	NACA 4421	NACA 4414	NACA 4412
Weight: Blades and hubs		40000 kg		
Shaft, housing and bearings		19500 kg		
Moment of inertia (about axis of rotation)		6.12×10^{6} kg m^{2}		

*chord line to plane of rotation

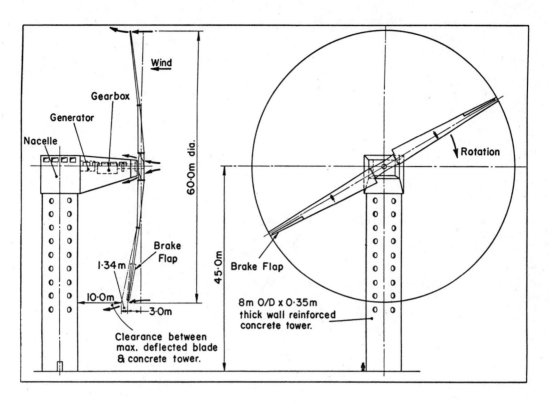

Fig. 1   General arrangement of concrete tower scheme

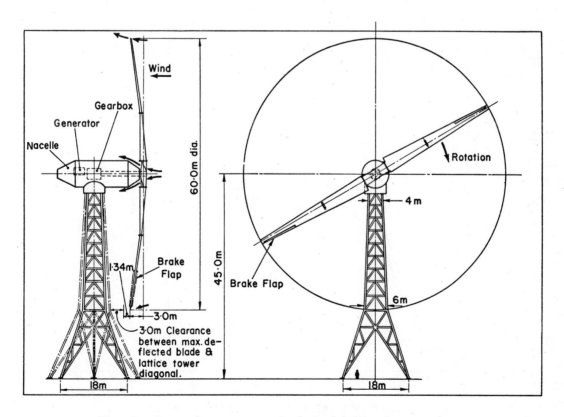

Fig. 2   General arrangement of steel lattice tower scheme

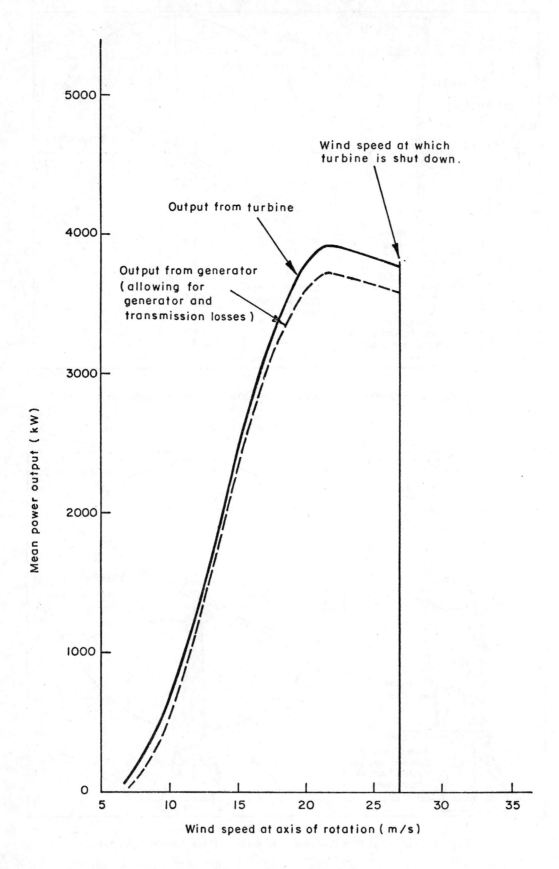

Fig. 3 Power - wind speed characteristic for
base-line design, 60m dia. wind turbine

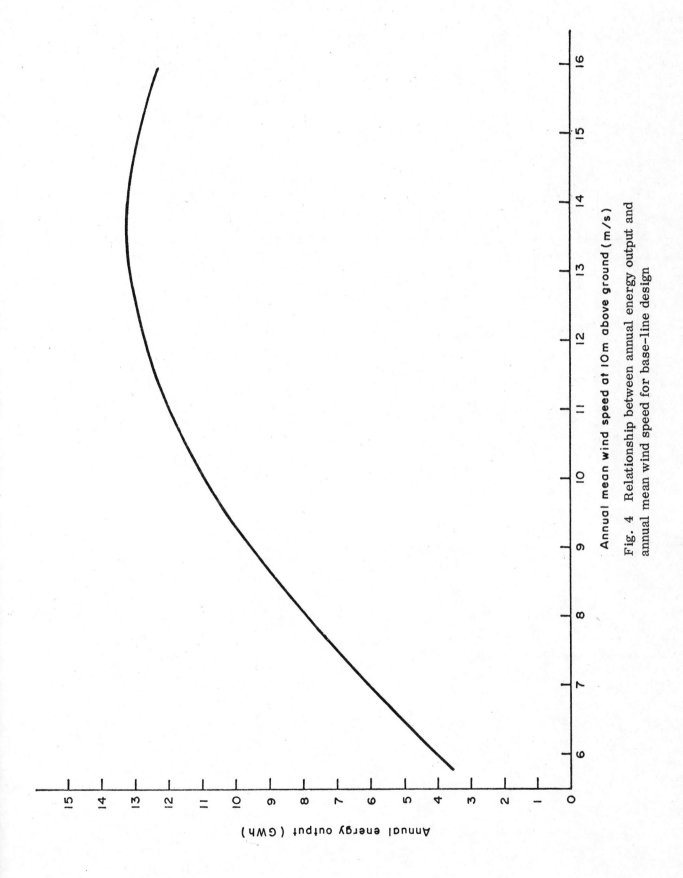

Annual mean wind speed at 10m above ground (m/s)

Fig. 4   Relationship between annual energy output and
annual mean wind speed for base-line design

WIND ENERGY SYSTEMS

October 3rd - 6th, 1978

THE CYCLOTURBINE AND ITS POTENTIAL
FOR BROAD APPLICATION

H.M. Drees

Pinson Energy Corporation, U.S.A.

Summary

The Cycloturbine, developed by Pinson Energy Corporation, a vertical axis, straight-bladed, wind driven turbine with cyclically pitched blades is presented. Its construction, and considerations used in the design are given in addition to the turbine's performance characteristics. The machine's versatility, which gives it a potential for broad application, as well as the Cycloturbine's present state of development is also described.

Held at the Royal Tropical Institute, Amsterdam, Netherlands.

Symposium organised by BHRA Fluid Engineering in conjunction with the Netherlands Energy Research Foundation ECN

©BHRA Fluid Engineering, Cranfield, Bedford, MK43 0AJ, England.

INTRODUCTION

The Cycloturbine is a vertical axis, straight-bladed wind driven turbine with cyclically pitched blades. (Fig. 1) Originally conceived and built at the Massachusetts Institute of Technology through a National Science Foundation grant, (Ref. 1) it is similar in aerodynamic operation to a design patented in 1931 by G.J.M. Darrieus. Pinson Energy Corporation was formed in 1975 to develop and produce the Cycloturbine. Pinson Energy has its offices in Marstons Mills, Massachusetts, and a test facility at New Seabury, both on windy Cape Cod. Testing of a protopype Cycloturbine was initiated early in 1976. The test program yielded data utilized in the design of an improved second prototype. Testing of this second machine is nearing completion and presently eight "pre-production" units are being installed as part of Pinson Energy Corporation's field test and demonstration program.

The Cycloturbine differs from the classic "eggbeater" rotor in that its blades do not remain at a fixed angle, but follow a preset schedule of angle change. The amount and timing of pitch change is determined by a cam device mounted atop the main shaft, actuating the blades via pullrods. (Fig. 2) A tailvane affixed to the cam determines correct orientation relative to the wind. The combination of cyclic pitching and straight blades have advantages, both mechnaical and economical, over the Darrieus design. The variably angled blades of the Cycloturbine allow it to self-start in low winds. They also allow for overspeed control through a centrifugally actuated modification of the blade pitch angle schedule. Specifications are found in Table 1.

CONSTRUCTION and Design Considerations

The blades of production Cycloturbines are straight and untwisted with a rectangular planform and are relatively inexpensive and simple in construction. They are constructed with a tubular spar, ribs, and an aluminum skin which form a NACA 0015 airfoil profile. The structure is riveted and epoxy-bonded together in assembly. Blade construction is further described in Ref. 2 & 3. In operation, the blades are subject to two types of loads, both are distributed. Centrifugal loads predominate under normal, low wind (13.5 m/s) conditions while oscillatory aerodynamic loading predominates under controlled conditions in high wind speeds.

The maximum Cycloturbine rotor speed is limited at 200 rpm. At this rotational speed, the centrifugal loading on the blades is 81 times the force of gravity. Centrifugal loading on the spar is dependent solely upon the rotational speed and the weight of the blade. It is, therefore, critically important to design and construct a light weight, yet stiff blade.

A Cycloturbine blade weighs 5 kg. The distributed centrifugal loading on the blade at a rotor speed of 200 rpm is 169 kg/m, corresponding to a blade weight of 405 kg at 81 g. If the blades are attached to the struts so as to optimize for minimum bending stress, the moment distribution shown in Figure 3 will result.

It can be deduced that a Cycloturbine should be designed to have a diameter not less than 3.6 m unless the solidity is increased to keep rotational speed, and therefore, centrifugal blade spar stress low since the centrifugal loads are inversely proportional to the diameter and proportional to the square of the rotational speed.

The six struts of the Cycloturbine transfer the forces imparted by the blades to the hubs on the main shaft. The primary spanwise loads on the struts are a result of aerodynamic and centrifugal blade loads, in

addition to the centrifugal force that the struts contribute themselves.
The maximum radial loads occur at the hub end of the struts. The struts,
however, must be designed to withstand oscillatory in-plane forces as
a primary concern. These forces are caused by the tangential aerodynamic loading and unloading of the blades and are transferred to the
hub as a moment producing a two/rev "lead-lag" stress at the root of
the struts. Since the resulting stress is oscillatory in nature, fatigue weakening of the strut can occur and the strut framework must be
designed to have a root moment of inertia (area) so that stresses are
kept below fatigue limits.

The construction of the struts on the Cycloturbine consists of two
aluminum rectangular tubes forming a narrow triangle with the base at
the hub plate connection. This construction meets the flatwise moment
of inertia criteria. The framework is strengthened by welded gusset
plates acting as structural doublers at each end. The struts are
cowled to airfoil shape. The blades are connected to the strut by
means of a hinge. The hinge consists of a bellcrank assembly welded
to the blades which attaches to the struts by means of a "hinge-pin"
bolt and Teflon bushings allowing the blade to pivot in pitch change.
The struts are connected to steel hub plates by a sandwich type construction so that the hub bolts carry loads in double shear. The hub plates
are welded to a steel mainshaft which is supported by the bearing cart-
ridge weldment that bolts to the tower top. The shaft is of sufficient
diameter so that its first bending mode natural frequency is well above
those frequencies imparted to it by the rotor.

Tests have shown that the shaft, and therefore, the tower, receive a
rotor-induced aerodynamic loading three times per revolution, probably
originating in a wake-blade interaction in the rear section of the
rotor. At a range of rotor speeds of 0 to 200 rpm, the shaft experiences
a loading with a frequency of 0 to 10 Hertz. It has been found that the
rotor/tower system must have a first mode natural frequency which is
lower than 20% of the rotor angular frequency in order that the rotor
runs super-critically to the tower's first mode frequency.

The drive-train is a 7.5:1 ratio, double-step, cogbelt type speed changer
which drives a 110/220 V 4 KW alternator.

The aerodynamic controls consist of a fulcrum balance which reacts to
centrifugal loads of the pullrods transmitted through the cam, and ad-
justs the cam position which in turn controls the blade pitch. The
adjusted pitching motion of the blades changes the aerodynamic operation
of the turbine and limits the rotational speed. At a windspeed of
18 m/sec, the turbine will automatically shut down and stop by means of
a trip mechanism.

PERFORMANCE

The Cycloturbine, tested at Pinson Energy's test site on the coast of
Cape Cod, has shown promising performance characteristics. A power
coefficient vs tip speed ratio data plot is given in Fig. 4. The data
were gathered by simultaneously measuring the turbine's rotational speed
and the wind speed at the center of the rotor three diameters upstream.
The electrical output was also recorded using a resistive load. It is
seen in the figure that a maximum C_p (electrical load) of about .45
occurs when λ is 3.5 under very smooth wind conditions. The run-away
tip-speed ratio is about 5. A "nominal" C_p of 0.3 is maintained in
gusty conditions.

In its operation, the Cycloturbine has shown that it is stable and
smooth-running. Aerodynamic noise is very low. The Cycloturbine was
designed with a solidity (blade planform area/projected rotor area) of

0.25 and is, therefore, a slow running machine (by wind turbine generator standards). This is advantageous in terms of wear and lifetime of cyclically loaded parts. The relatively high solidity also results in a high starting torque and low windspeed startup.

VERSATILITY IN APPLICATION

Because of its vertical axis of rotation, the Cycloturbine has several advantages over horizontal axis propeller-type wind turbines. The Cycloturbine is omnidirectional in its acceptance of the wind. Sudden changes in wind direction are reacted to rapidly by the lightweight tailvane, keeping the cam axis oriented into the wind. A propeller-type wind turbine, on the other hand, must be oriented in its entirety by a large trailing vane, or by virtue of downwind rotor placement. Because most propeller-type machines are mounted directly to their attendant generating device atop their towers, they are relatively massive in form, and cannot orient as responsively as the Cycloturbine. Additionally, the Cycloturbine does not experience gyroscopically induced stresses found in horizontal, propeller-type wind generators because orientation takes place in the plane of rotation.

The most useful advantage of the vertical configuration of the Cycloturbine is the versatility and simplicity afforded by the vertical main shaft in mounting driven devices. A variety of driven devices, such as alternators, generators, compressors, pumps, or fluid friction heaters can be placed below the rotor and driven by the working shaft. The mass of the working device is low and out of the air flow through the rotor. The Cycloturbine will accept any such device with only mount and transmission changes, whereas most propeller-type machines are designed around a specific driven device, limiting their use to one purpose. In addition, no slip-rings are needed. Finally, should a change or repair on driven device be necessary, the entire rotor need not be removed; in fact, it is possible to have the driven device at or near ground level, greatly facilitating access through the use of a bearing supported shaft extending to the ground. In short, the Cycloturbine's mechanical and aerodynamic design characteristics, in combination with its vertical axis configuration, gives the turbine a potential for broad application which is superior to that of other wind turbine configurations.

CONCLUSION and Present Cycloturbine Development

In an effort to demonstrate the potential for broad application of the Cycloturbine, Pinson Energy Corporation is collaborating with universities, institutions, and homeowners by installing turbine systems for a number of different uses. Among those presently under installation and their respective end uses are:

- Cornell University for driving a device which heats water through friction by stirring. This is a United States Department of Agriculture funded research project.

- New Alchemy Institute on Cape Cod for providing mechanical power to run an air compressor. Compressed air is used for various tasks in a food producing greenhouse and fish breeding tanks.

- Rockwell International Rocky Flatts Plant in Boulder, Colorado (Under contract by the U.S. Department of Energy) as an instrumented experimental machine for evaluation.

- Homeowners installation for providing resistive heat for water heating and other electrical power for use in homes.

Performance data will be taken for each installation and evaluated as to the practicality. These data will be available in future publications.

Pinson Energy Corporation is presently under contract by the US Department of Energy/Rockwell International in conjunction with Aerospace Systems Incorporated of Burlington, Massachusetts in a two year program to "Develop a 2 KW High Reliability Cycloturbine" to be completed by the end of 1979.

It is expected that due to its versatility and simplicity in construction that the Cycloturbine will be a practical solution to at least small scale conversion and utilization of wind energy.

REFERENCES

1. Drees, H.M. "An Analytical and Experimental Investigation of a Darrieus Wind Turbine Employing Cyclic Blade Pitch Variation". MS Thesis, Massachusetts Institute of Technology, Cambridge, Massachusetts, 1976.

2. Drees, H.M. "Blade Design and Construction of the Pinson Cycloturbine". 1978 Wind Turbine Blade Workshop Summary Proceedings, University of Massachusetts, Amherst, Massachusetts, January 1978.

3. Drees, H.M. "Design Considerations of the Cycloturbine". American Wind Energy Association Conference Proceedings, Amarillo, Texas, March 1978.

SPECIFICATIONS OF THE CYCLOTURBINE

Length of Blades:	2.4 m
Chord of Blades:	29 m
Type of Airfoil:	NACA 0015
Diameter of Rotor:	3.6 m
Weight of Rotor:	50.5 kg
Type of Overspeed Control:	Centrifugally Activated
Starting Windspeed:	2.2 m/s
Governing Windspeed:	13.5 m/s
Governing RPM:	200
Shut-down Windspeed:	20 m/s
Electrical Output: (w/4KW Winco Alternator)	2KW @ 11m/s 4KW @ 13.5 m/s

TABLE 1.

FIGURE **1**
PINSON ENERGY CORPORATION CYCLOTURBINE
Model C2E

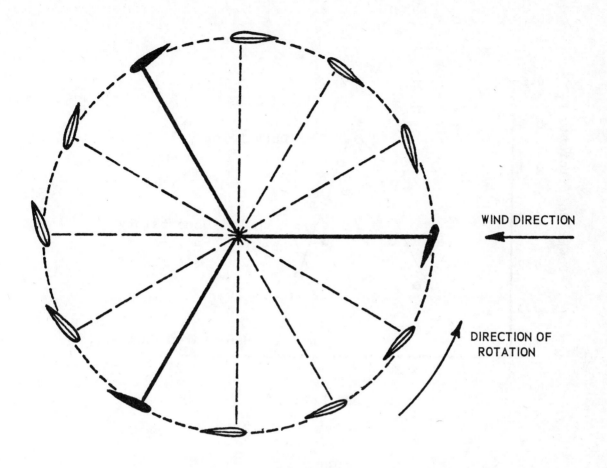

WIND DIRECTION

DIRECTION OF
ROTATION

FIGURE 2
A REPRESENTATION OF THE OPERATION OF THE CYCLOTURBINE

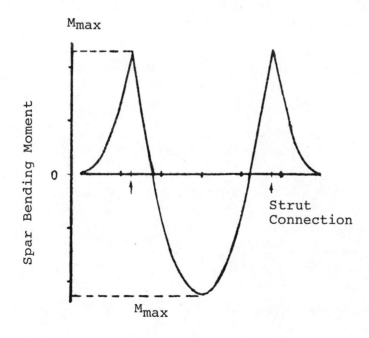

M_{max}

Spar Bending Moment

0

Strut
Connection

M_{max}

FIGURE 3
SPAR BENDING MOMENT DISTRIBUTION
DUE TO CENTRIFUGAL LOAD

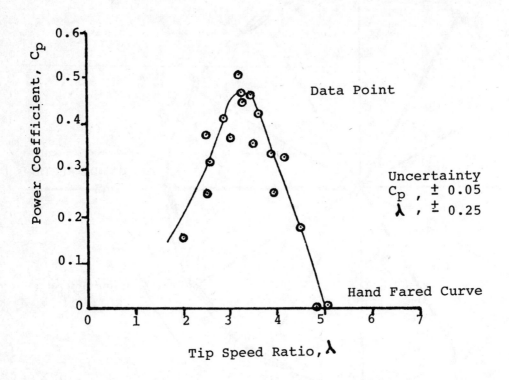

FIGURE 4
PERFORMANCE ENVELOPE FOR THE CYCLOTURBINE

A COMPARISON OF AERODYNAMIC ANALYSES FOR THE DARRIEUS ROTOR

R.E. Wilson and W.R. McKie

Oregon State University, U.S.A.

Summary

A comparison is made of the single streamtube, multiple streamtube, fixed wake and free wake analyses of a straight bladed Darrieus Rotor using potential flow aerodynamics. The angle of attack, lift coefficient, circulation and loads are examined for a rotor operating at maximum performance. The unsteady aerodynamic forces are evaluated with special consideration given to wake crossing transients.

Held at the Royal Tropical Institute, Amsterdam, Netherlands.

Symposium organised by BHRA Fluid Engineering in conjunction with the Netherlands Energy Research Foundation ECN

NOMENCLATURE

c   Blade chord

C_L Blade lift coefficient

C_M Blade moment coefficient

C_x Windwise direction force coefficient

C_Y Crosswind direction force coefficient

C_P Power coefficient

F_R Blade radial force

F_t Blade tangential force

R   Rotor radius

U   Freestream wind speed

Δu Wake induced x-direction velocity at the blade

Δv Wake induced y-direction velocity at the blade

W   Relative velocity at the blade

X   Tip speed ratio

x   Windwise coordinate

y   Crosswind coordinate

α Angle of attack

β Blade angular position

Γ Circulation on the blade

Ω Angular velocity of blade

INTRODUCTION

Although invented in 1926, [1] the Darrieus Rotor did not see extensive development until the 1970's when South and Rangi [2,3] and later the Sandia Laboratories [4] undertook the analysis, design and construction of large Darrieus Rotors. In the course of this development, many aerodynamic models for the Darrieus Rotor have been proposed. These models vary considerably in their treatment of the flow with no single approach available that covers stall, variable induced flow, unsteady pitching moment and wake crossing. In this paper a comparison of the predictions of four methods will be made in order to gauge their strengths and weaknesses.

PREVIOUS STUDIES

Flow models of the Darrieus Rotor have been published by Templin [5], Wilson and Lissaman [6], James [7], Muraca [8], Shankar [9], Strickland [10], Holme [11], Fanucci and Walters [12], and Ashley [13]. These models may be classified into three groups:

a. Streamtube models [6,5,8,9,10].

b. Fixed-wake models [11].

c. Free-vortex models [12].

STREAMTUBE MODELS

The streamtube models calculate the induced axial velocity at the rotor by equating the time-averaged force on the blades to the mean momentum flux through a streamtube of fixed location and dimensions. In Templin's approach a single streamtube is used for the entire rotor; in the Wilson-Lissaman approach, multiple streamtubes are used. Both methods predict fore and aft symmetry of flow quantities due to the fact that the streamtubes used have constant dimensions through the rotor.

The forces are calculated from the local angle of attack and local velocity using static airfoil data for lift and drag coefficients as a function of angle of attack. The streamtube approaches have met with modest success in predicting the overall performance and axial force of Darrieus Rotors.

The streamtube approaches can treat curved bladed rotors and can incorporate any variation of lift coefficient with angle of attack. Additionally, the computation time needed for these methods is much less than any other approach.

The streamtube approaches do have several shortcomings and the agreement of streamtube predictions with wind tunnel tests warrants some caution. First, the wake induced velocities at the rotor are known to be in error. Another of the shortcomings of these approaches lies in the quasi-static aerodynamics which neglects the effects of the unsteady wake, the pitching circulation and wake crossing. Consequently, although the mean loads are in general agreement with experiment, the local aerodynamic loads predicted by streamtube methods are known to be in error.

Finally, the airfoil data selected to make analytic predictions warrant caution. The integrated values for torque and axial force are quite sensitive to the lift curve slope, drag coefficient and stall angle. Relatively small changes in static aerodynamics can produce noticeable shifts in the calculated power and force.

FIXED WAKE MODEL

The fixed wake model was developed by Holme [11] and has been extended by the authors [14] to yield the flow field upstream and downstream of the rotor. In this approach, the rotor contains an infinite number of infinitesimal blades distributed around the turbine in such a manner that the term Nc/R maintains a fixed value. Here N is the number of blades and c is the blade chord so the Nc/R is the ratio of total chord length of blades to the rotor radius. The vortex field of the flow consists of a sheet bound to the rotor and wake sheets shed from the rotor in the direction of the free stream. Both the bound and wake sheets are independent of time. The shed wake moves at a uniform speed.

The forces are determined from the Kutta-Joukowski Law using the local relative velocity on the rotor and the local circulation. Figure 1 shows the streamlines and velocity profiles for a Darrieus Rotor operating at a tip speed ratio of 3.54 with Nc/R=0.2. This is a maximum power condition. The fixed wake analysis has several advantages over the streamtube methods and the cost to make the calculations is moderate, on the same order as the streamtube approaches. The advantages in prediction stem from the ability to treat the differences between the upwind and downwind blade loadings. There are, however, shortcomings of this approach. These are:

1. The method has not been developed to treat curved blades. This can be accomplished.

2. The method uses $C_L = 2\pi\sin\alpha$ so that blade stall is not covered. With curved bladed Darrieus Rotors, the innermost portions of the blades operate at high angles of attack and are very likely to be stalled.

3. The approach does not treat the unsteady aerodynamic loads.

FREE-VORTEX MODEL

The most complex method of analysis that we have developed [15] for the Darrieus Rotor is the free-vortex approach. Using the circle theorem, the airfoil is mapped into the circle plane where the Kutta Condition is satisfied and the strength of the induced center, shed and image vortices are determined. The wake is modelled by discrete, force-free vortices convected downstream with velocity determined by induced velocities from the rest of the·system. The method has been checked by comparison with the Kármán-Sears solution for a plunging flat plate and the discrete vortex approach yields excellent agreement with the analytical solution.

The forces and the moment on the blade are determined by two independent methods:

1. Integration of the pressure over the airfoil using the unsteady Blasius theorem;

2. Determination of the impulse of the wake vortices.

Agreement between these two approaches to two significant figures indicates that we indeed have a force-free wake.

Figure 2 illustrates the shed wake vortex sheet location for a heavily loaded Darrieus Rotor at a tip speed ratio of 2. The sheet location was determined (with artistic license) by drawing connecting lines between consecutive shed point vortices. The Kármán-Vortex-Street-like appearance of the wake is to be noted.

The free-vortex approach stands to yield understanding of the role of unsteady aerodynamics for the Darrieus Rotor. Both the unsteady flow and the wake crossing are found to have a significant role in determining the forces on the Darrieus Rotor blades. Additionally, the moment is determined by this approach.

The free-vortex approach, however, has several disadvantages. These are:

1. It is quite expensive to use.

2. The effects of stall are not included.

3. At the present, only a single, straight-bladed rotor may be treated.

4. The values for the performance and loads are approached asymptotically. Since the wake never reaches an infinite length, the loads and performance are higher with a finite length wake than would be obtained with an infinite wake. The values presented in this paper are estimated to be within 4% of the final asymptotic values.

MODEL FOR COMPARISON OF METHODS

In order to compare the results of the various approaches, the following case has been selected. The geometry is illustrated in Fig. 3.

$$\frac{c}{R} \equiv \frac{\text{Rotor chord}}{\text{Rotor radius}} = 0.2$$

$$\frac{R\Omega}{U} \equiv \frac{\text{Tip Speed}}{\text{Wind Speed}} = 3.54$$

$$N \equiv \text{Number of Blades} = 1$$

The case listed above corresponds to a heavily loaded Darrieus Rotor. Table 1 gives the overall performance parameters calculated by each method. It may be noted that the power coefficient and windwise force coefficient for all methods to be very close together.

COMPARISON

The angle of attack history for a rotor going through a cycle is shown in Fig. 4. Again the agreement between the various methods is good with the exception that the streamtube methods fail to predict the fore and aft angle of attack differences. At a tip speed ratio of 3.54 this fore and aft difference is in the range of 2 to 4 degrees. In an actual rotor, the blades in the upwind position will experience higher than predicted angles of attack when using streamtube methods.

Figure 5 illustrates the variation of lift coefficient during a cycle for the Darrieus Rotor. This figure illustrates that there is a significant contribution from the unsteady terms. Whereas the angle attack of variation between the various methods was relatively small, the lift coefficient shows a peak value 50% greater than the value of C_L predicted by the multiple streamtube method.

The effects of the unsteady aerodynamics are illustrated in a different fashion in Fig. 6 where the lift coefficient is plotted against the angle of attack for two different values of chord to radius ratio. The plot shows that at a tip speed ratio of 3.54 the lift coefficient leads the angle of attack by an amount approximately equal to 15 c/R degrees. The variation in lift coefficient during the wake crossing is also shown for case where c/R = 0.2.

The circulation is shown in Fig. 7. Both the difference in magnitudes and the phase shift of the circulation may be noted.

The radial and tangential loads on the rotor are illustrated in Figs. 8 and 9. The radial load merits particular attention since the free-vortex approach indicates peak radial loads from 22% to 50% higher than the other methods. The in-plane (tangential) force also indicates the lack of fore-aft symmetry. In both of these figures the forces predicted by the free vortex method shows a fluctuation in force near $\beta = 320°$. This is due to wake crossing effect. Since there is only one blade for this model, there are fewer wake crossings possible than for the case of a multi-bladed rotor. Figure 10 illustrates the angle of attack and force variation during the blade crossing the wake near $\beta = 320°$. The individual points are plotted rather than curves drawn since the computational step size (1° here) determines the frequency resolution.

The pitching moment, (positive counter clockwise) is shown in Fig. 11 for chord to radius ratios of 0.1 and 0.2. The wake crossing effect is detailed for the case of the larger chord. The effect of the unsteady pitching moment on the power is small, less than 1% in both cases.

Since the maximum moment coefficient varies as $X(c/R)^2$ and loading, the moment becomes quite small for multiblade rotors. Note that the maximum pitching moment for the c/R = 0.1 case is approximately 1/4 of the c/R = 0.2 case.

Figures 12 and 13 illustrate the induced x and y direction velocities predicted by the four different approaches. The lack of fore and aft symmetry is evident in Fig. 12

where the windwise induced velocity is shown. The crosswind (y-direction) wake induced velocity is illustrated in Fig. 13. The effect of having only a single blade in
the free vortex analysis is very evident here. Whereas the fixed wake analysis pre-
dicts crosswind velocities that are positive in the right hand half of the rotor
($\pi/2 > \beta > -\pi/2$) and negative crosswind velocities in the left-hand half of the rotor,
the free vortex approach with a single blade shows positive crosswind velocity for all
blade positions.

The reason for this behavior is straightforward. The shed vorticity for a single blade
rotor shown in Fig. 2 is counterclockwise immediately behind the blade. This induces a
positive y-direction velocity on the blade. In the fixed wake analysis the induced
velocity from the wake for a blade in the position illustrated in Fig. 2 is in the negative y-direction. The shed vorticity in the left-hand half of the rotor is again
counterclockwise, however the majority of the shed vorticity for the fixed wake case is
downwind of the blade and hence it induces a crosswind velocity opposite that of the
vorticity near the trailing edge.

DISCUSSION

There are large differences in loads predicted by the various approaches to the
Darrieus aerodynamics.

The overall performance parameters, however, such as power and force coefficients,
are in general agreement. Part of the differences predicted by the free vortex
approach may be attributed to the fact that a single blade was used and thereby the
pitching and apparent mass terms make a larger contribution than if several blades
were used. Additionally, the indiced velocities are different, particularly the local
crosswind induced velocities.

The effect on the loads, however, shows that the pitching term make very little contribution to the radial load, although a significant contribution to the maximum torque.
The apparent mass term on the other hand makes no torque contribution and is out of
phase with the overall radial load. The major difficulties with the streamtube approaches lie in the facts that -

1. They do not predict the variation of induced velocity between the front and rear
    of the rotor. Hence radial loads are larger and blade stall will occur sooner
    than predicted by the streamtube approaches.

2. The lift coefficient leads the angle of attack.

The fixed wake analysis also appears to give lower than actual loads, however the fact
that a single blade is used in the free wake analysis accounts for part of the difference. Significant contributors to the unsteady aerodynamics are -

1. The pitching circulation and the apparent mass both of which are proportional
    to $X(c/R)^2$. Since maximum power occurs for the Darrieus at $XNc/R = 0.8$, if the
    number of blades is increased, the magnitude of these terms relative to the
    Kutta-Joukowski term will decrease as $1/N$ at maximum power.

2. The unsteady circulation term, $c/2R \ X \ \partial/\partial\beta \ (\Gamma/UR)$ which at maximum power will
    decrease as $1/N$ relative to the Kutta-Joukowski term.

3. The unsteady moment term which will also decrease as $1/N$ at maximum power.

CONCLUSIONS

There are significant differences in the loads predicted by the various methods of
analysis of the Darrieus Rotor. The overall performance, in the absence of stall,
appears to be relatively insensitive to the method of analysis. The differences in
induced velocities and the phase shift in lift coefficient are the major effects in
causing the loads predicted by the streamtube methods to be in error.

Several ad-hoc modifications in the available models for Darrieus aerodynamics are possible in order to improve the ability to predict loads and performance. In view of the large crosswind induced velocities predicted by the fixed wake analysis under conditions of heavy loading, modifications to the streamtube analysis do not appear to have great potential. An approach whereby the induced velocities are predicted by a fixed wake analysis and then the blade loads determined by quasi-steady aerodynamics is recommended for further development.

REFERENCES

[1] Darrieus, G.J.M.: U.S. Patent No. 1,835,018, December 8, 1931.

[2] South, P. and Rangi, R.: "The Performance and Economics of the Vertical-Axis Wind Turbine Developed at the National Research Council, Ottawa, Canada," presented at the 1973 Annual Meeting of the Pacific Northwest Region of the American Society of Agricultural Engineers, Calgary, Alberta, Oct. 19-12, 1973.

[3] South, P. and Rangi, R.S.: "An Experimental Investigation of a 12 Ft. Diameter High Speed Vertical Axis Wind Turbine," National Research Council of Canada, LTR-LA-166, April 1976.

[4] Blackwell, B.F. and Reis, G.E.: "Blade Shape for a Troposkien Type of Vertical-Axis Wind Turbine," Sandia Laboratories, Albuquerque, N.M. SLA-74-0154, April 1974.

[5] Templin, R.J.: "Aerodynamic Performance Theory for the NRC Vertical-Axis Wind Turbine," National Research Council of Canada, LTR-160, June 1974.

[6] Wilson, R.E. and Lissaman, Peter B.S.: "Applied Aerodynamics of Wind Powered Machines," Oregon State University, May 1974.

[7] James, E.C.: "Unsteady Aerodynamics of Variable Pitch Vertical Axis Windmill," AIAA Paper No. 75-649, AIAA/AAS Solar Energy for Earth Conference, Los Angeles, CA, April 21-24, 1975.

[8] Muraca, Ralph J., Stephen, S., Maria, V., and Dagenhart, Ray J.: Theoretical Performance of Vertical Axis Windmills," NASA-Langley Research Center, NASA TM TMX-72662, May 1975.

[9] Shankar, P.N.: "On the Aerodynamic Performance of a Class of Vertical Axis Windmills," National Aeronautical Laboratory, Bangalore TM AE-TM-13-75, July 1975.

[10] Strickland, J.H.: "The Darrieus Turbine: A Performance Prediction Model Using Multiple Streamtubes," Advanced Energy Projects Department, Sandia Laboratory, SAND 75-0431, October 1975.

[11] Holme, Olof: "A Contribution to the Aerodynamic Theory of the Vertical-Axis Wind Turbine," International Symposium on Wind Energy Systems, Cambridge, England, September 1976.

[12] Fanucci, J.B. and Walters, R.E.: "Innovative Wind Machines: The Theoretical Performance of a Vertical Axis Wind Turbine," p. III-61-95, Proceedings of Vertical-Axis Wind Turbine Technology Workshop, Sandia Laboratory, Albuquerque, N.M., SAND 76-5586, May 1976.

[13] Ashley, Holt: "Some Contribution to Aerodynamic Theory for Vertical Axis Wind Turbines," 12th IECEC, Washington, D.C. 28 August, 1977.

[14] McKie, W.R., Wilson, R.E., and Lissaman, P.B.S.: "The Internal and Local Flow Fields of Vertical Axis Wind Turbines," Paper submitted to the AIAA Journal of Energy.

[15] Lissaman, P.B.S., Wilson, R.E., and James, M.: "Calculation of Forces and Moments on an Airfoil in Non-Steady Motion Using Discrete Vortices," paper in preparation.

Table 1. Overall Performance Predictions For a

Darrieus Rotor NC/R = 0.2, X = 3.54

Method	C_P	C_x	C_y
Single Streamtube	0.580	0.803	0
Multiple Streamtube	0.548	0.778	0
Fixed Wake	0.541	0.746	0.048
Free Wake			
(a) for the 8th revolution	0.569	0.782	0.068
(b) for the 9th revolution	0.560	0.760	0.065
(c) for the 10th revolution	0.554	0.772	0.064
(d) asymptotic projection	0.544	0.764	0.063

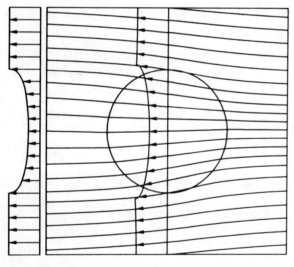

Fig. 1   Flow field about a heavily loaded Darrieus Rotor computed by the fixed wake analysis. The downstream profile is 10 radi from the rotor centre. Tip speed ratio is 3.54, Nc/R = 0.2.

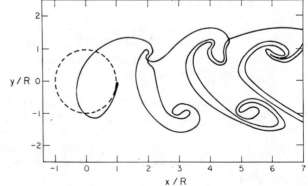

Fig. 2   Wake vortex sheet of a heavilyloaded one-bladed Darrieus Rotor. Tip speed ratio is 2, c/R = 0.2.

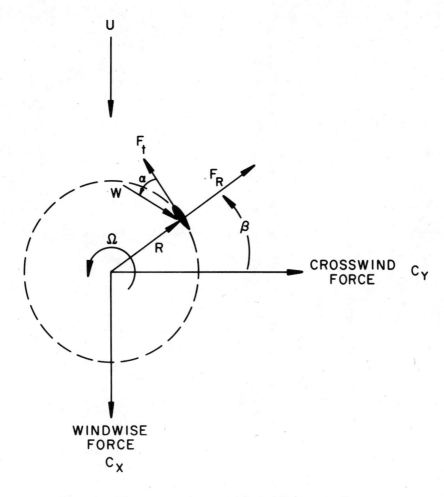

Fig. 3   Plan view of a one-bladed Darrieus Rotor.

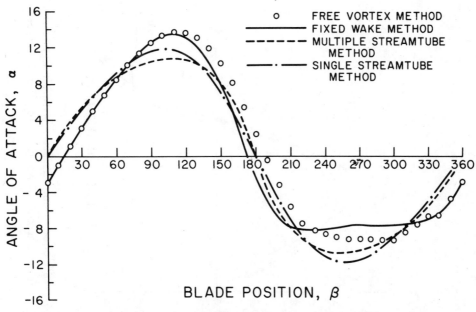

Fig. 4   Angle of attack variation with blade position using four methods of analysis for the Darrieus Rotor.

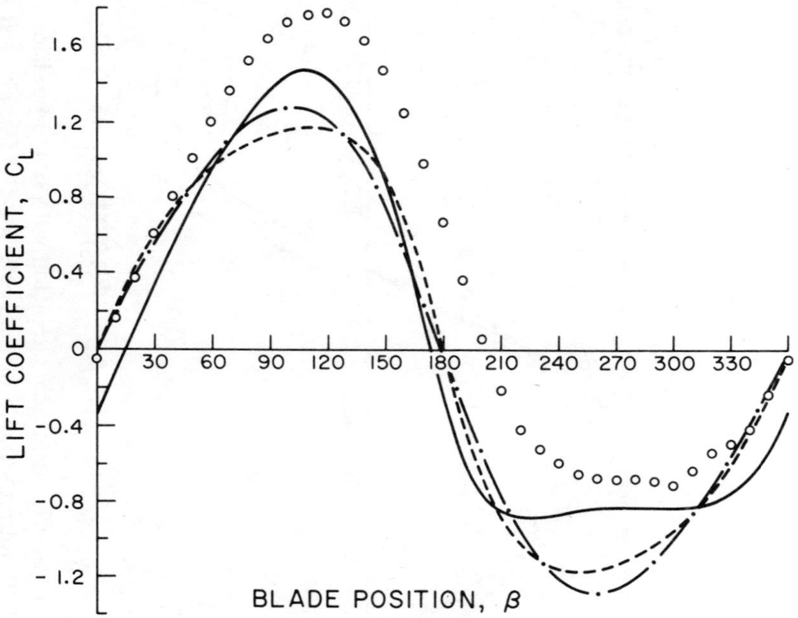

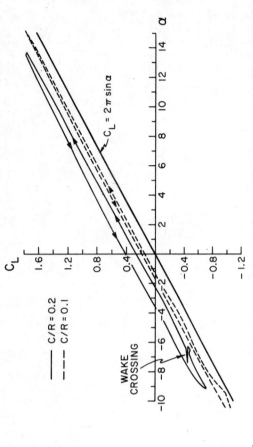

Fig. 6 Lift coefficient variation with angle of attack for a one-bladed Darrieus Rotor using the force-free analysis. Tip speed is 3.54, c/R = 0.1 and 0.2.

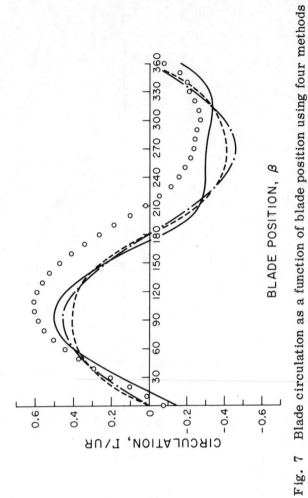

Fig. 7 Blade circulation as a function of blade position using four methods of analysis for the Darrieus Rotor. X = 3.54, c/R = 0.2.

Fig. 5 Lift coefficient variation with blade position using four methods of analysis for the Darrieus Rotor.

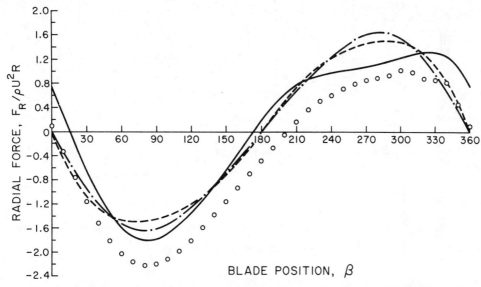

Fig. 8 Blade radial force as a function of blade position using four methods of analysis of the Darrieus Rotor. X = 3.54, c/R = 0.2.

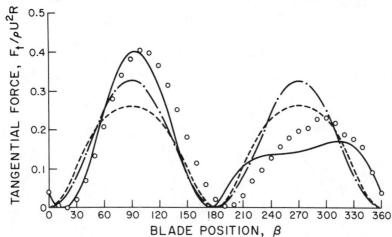

Fig. 9 Blade tangential force as a function of blade position using four methods of analysis of the Darrieus Rotor. X = 3.54, c/R = 0.2.

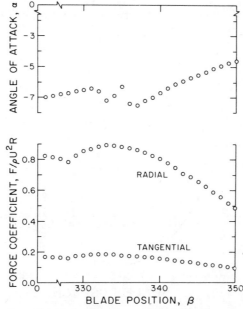

Fig. 10 Blade angle of attack history and loading history during a wake crossing for a one-bladed Darrieus Rotor. X = 3.54, c/R = 0.2.

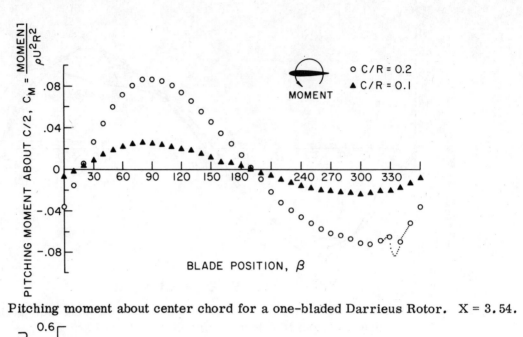

Fig. 11   Pitching moment about center chord for a one-bladed Darrieus Rotor.   X = 3.54.

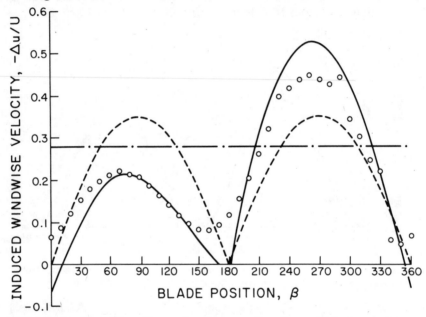

Fig. 12   Wake induced x-direction velocity as a function of blade position.   X = 3.54.

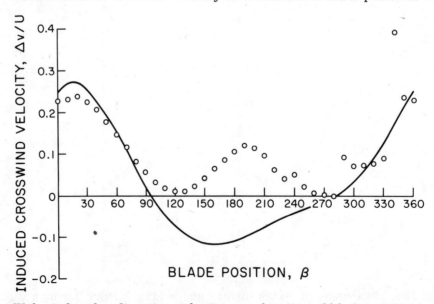

Fig. 13   Wake induced y-direction velocity as a function of blade position.   X = 3.54.

WIND ENERGY SYSTEMS

October 3rd - 6th, 1978

A THEORETICAL AND EXPERIMENTAL INVESTIGATION
INTO THE VARIABLE PITCH VERTICAL AXIS WIND TURBINE

W. Grylls, B. Dale and P.-E. Sarre

University of Exeter, U.K.

Summary

A computer model of the variable pitch vertical axis wind turbine is used to investigate the power characteristics of this class of machine. The paper goes on to describe the construction and testing of a 2.4 m diameter prototype. The correlations between the experimental and theoretical findings are discussed.

Held at the Royal Tropical Institute, Amsterdam, Netherlands.

Symposium organised by BHRA Fluid Engineering in conjunction with the
Netherlands Energy Research Foundation ECN

NOMENCLATURE

α	angle of attack
γ	angle of blade centre line to tangent of blade locus
$\dot{\gamma}$	rate of change of γ
θ	position of rotor blade relative to datum Fig. 1
μ	coefficient of friction of blade root bearing
ρ	density of air
ρ_{al}	density of aluminium alloy
a	half amplitude pitch variation
A	Aspect Ratio of the blades Blade length/chord
c	chord of aerofoil
C_D	coefficient of drag
C_L	coefficient of lift
C_p	Power coefficient: Power extracted from the wind by rotor/ total power in wind (same area as rotor).
$\Delta C_{p_{aero}}$	Power Coefficient loss due to aerodynamic drag
$\Delta C_{p_{fric}}$	Power Coefficient loss due to bearing friction
h_1 & h_2	coefficients describing radial and thrust loads on blade root bearings
H	Overall height of rotor blades
K_A	Coefficient describing power coefficient loss due to aerodynamic drag
K_F	Coefficient describing power coefficient loss due to bearing friction
N	Number of blades
r	a radius
$\bar{r}$	mean radius at which friction in bearing acts
R	Radius of rotor
S	Solidity of Rotor N c/R
T_D	Torque due to aerodynamic drag
T_F	Torque due to bearing friction
w	rotational speed of rotor
X	Tipspeed ratio Rotational Speed of blade/Speed of wind

1. INTRODUCTION

Design work began in late 1975 on a vertical axis wind turbine in the Department of Chemical Engineering. It was decided, in order to make the rotor self-starting and to obtain a power coefficient comparable to a horizontal axis rotor, to incorporate a mechanism for varying the position of each blade about its own axis (Fig. 1.). This type of system was discussed by Brulle and Larsen in June 1975 [1] and was also under development by Herman Drees at MIT (later with Pinson Energy Corporation).

The rotor and tower were constructed during 1976. Tests in the 10ft x 12ft working section of the closed return flow wind tunnel in the School of Engineering at the University of Bath, were carried out during 1977. These tests were to obtain an experimental power coefficient and to compare this with the value predicted from a computer model of the rotor.

2. COMPUTER MODEL

The multiple streamtube theory described by Strickland [2] was the basis of the computer model that describes the characteristics of the rotor. This theory has been simplified so as to incorporate the straight blades of the rotor (Fig. 2). It has been adapted to allow for a predetermined movement of the blades about their own axis in each cycle of rotor revolution (Fig. 1). This theory has been used in the past to predict the performance of a curved blade (troposkein) vertical axis rotor and has been shown to give a good correlation between theory and experiment [2,3]. In addition it is a more simple theory than the vortex theories, to adapt for use on the PDP 11 computer, which is available in the department.

The model allows for a variety of turbine characteristics which are input at the beginning of any computer run. These parameters are listed in Table 1. Two dimensional airfoil data is used, though corrections can readily be made to allow for finite aspect ratio. The corrections to the section drag coefficient and the angle of attack are:

$$C_D = C_{DO} + \frac{C_L^2}{\pi A} \tag{1}$$

$$\alpha = \alpha_o + \frac{C_L}{\pi A} \tag{2}$$

In studying the effect of various parameters such as solidity

$(S = \frac{Nc}{R})$ amount and type of pitch variation in a cycle and the mean angle of pitch, no allowance has been made for the aspect ratio as comparative values were all that were required in the initial study. In this study airfoil data taken from Sheldahl and Blackwell [4] has been used. The airfoil data used was for a NACA 0015 airfoil at a Reynolds number of 360,000.

We have looked through a wide range of solidity, the lowest being S = 0·05 and the highest being S = 0·9. In the computer model the number of blades is not an independent parameter as only solidity S is utilised in the equations. A more complete description of the programme can be found in [5].

Figure 3 illustrates the variation in the maximum power coefficient $(C_{p\ max})$ as the solidity (S) of the rotor is increased. The graph is for two types of straight bladed vertical axis turbines, firstly where

the pitch is varied sinusoidally in a cycle and secondly when there is
no variable pitch. This shows that when the blades are not moved during
the cycle there is a definite maximum power coefficient of 0·425 at a
solidity of 0·25. However with optimum sinusoidal pitch variation of
the blades there is not really an optimum solidity, a high power coefficient (> 0·425) can be obtained over a large range of solidity.

Fig. 4 shows how the tipspeed ratio (X) at the maximum power coefficient ($C_{p\ max}$) varies as a function of solidity. Overall as the
solidity increases so the tipspeed ratio at $C_{p\ max}$ decreases. This
curve appears to have an asymptote at $X \approx 2$. Also indicated in Fig. 4
is the amount of sinusoidal pitch variation required in a cycle to give
the maximum power coefficient at a given solidity. Here it shows how
as the solidity increases so the amount of pitch variation must be
increased to maintain the $C_{p\ max}$.

The variation in power characteristics as a function of the amount
of pitch for a rotor of solidity S = 0·3625 is shown in Fig. 5. After
reaching an optimum power coefficient at 3° pitch variation the $C_{p\ max}$
falls sharply and with a maximum pitch of 40° the power coefficient is
never positive. The location of the maximum pitch appears to be
insensitive to position. If it occurs between θ = 80 and θ = 120
degrees the change from the maximum C_p is less than 1%.

The variation of maximum pitch could be seen as a way of regulat-
ing the speed and power output of the rotor above a rated wind speed.
But another way is to change the angular position about which the blade
oscillates in each cycle. In Fig. 6 the effects of varying the average
pitch are shown for a turbine of solidity = 0·3625. With a mean pitch
angle of 27° the power coefficient is zero or less for any tipspeed
ratio.

A modification of the computer model which allowed the position of
the blade to be selected to give a maximum positive torque at each location in a blade cycle gave only a 2% improvement over the maximum power
coefficient obtainable with an optimum sinusoidal pitch variation. This
exercise was performed at a tipspeed ratio of 3 and a solidity of 0·3625.

3. ROTOR SYSTEM DESCRIPTION (Rotor Specification is listed in Table 2)

The blades of the rotor are fabricated from aluminium alloy tube
and sheet. The tube has hardwood formers fixed to it and the alloy
sheet is bent and glued to the formers and seam welded along the trailing
edge. The blades fit and are clamped into blade roots of stainless
steel (Fig. 7). The blade root is supported in a molybdenum disulphide
impregnated nylon bush. The blade root has a bell crank welded to it
and a rod end bearing is bolted to this. A rod then passes from the rod
end bearing through the support arm to a lever where the amount of pitch
is preset (Fig. 8). From here the mechanism is attached to the
eccentric bearing at the centre of the rotor. The rod between the bell
crank and the lever system is adjustable to set predetermined values of
average pitch.

The support arm is fabricated from aluminium alloy rectangular tube
and is covered in alloy sheet. At the blade root end of this arm fibre-
glass mouldings have been made to permit access to the bell-crank. The
support arms are bolted to the main hub. The rotor shaft is supported
by a set of tapered roller bearings and at the lower end a drum brake is
mounted. The rotor is mounted on a 4 m high octahedron module tower.
During wind-tunnel tests the shaft is extended from the brake via
universal joints to a friction drum that can be loaded using a leather
belt and weights (Fig. 9). The speed of the rotor is read from a hand

held tacheometer positioned at the lower end of the shaft

4. EXPERIMENTAL PROCEDURE

The floor of the 10ft x 12ft working section in the wind tunnel
(Fig. 10) has a circular hole cut in the steel floor and this is covered
with 25 mm blockboard. The centre board was removed and the VAWT was
mounted directly to the concrete floor approx. 1·1 m below the wind
tunnel floor. The dynamometer arrangement is shown in Fig. 9 and has
been described earlier. The wind speed in the tunnel was measured via
a series of kinetic and total head pressure probes, mounted on a stream-
lined rake 4·5 m upstream from the centre of the wind rotor. These
probes were connected to a multiple inclining manometer. Two of these
probes, at the height of the centre of the wind rotor were also connected to a Betz manometer to give accurate (± 0·02 mm of water) readings
of kinetic head. All these enabled us to see the vertical wind velocity
profile in the centre line of the tunnel, and an accurate reading of the
wind speed at the centre of the turbine to be taken.

When the rotor had been prepared for a run; after the amount of
variable pitch   the amount of average pitch and the location of the
maximum pitch had been set;   the wind tunnel was turned on and the
rotor was allowed to spin up to its runaway speed. Then while the
readings of the wind speed from the Betz manometer were constantly
monitored the rotor was slowly loaded up using weights. The values of
these weights together with the rotational speed of the rotor and the
windspeed were recorded when the rotational speed was observed to be
steady. As the weight difference was increased so the rotor slowed
down, until another steady rotational speed was observed, further sets
of readings were taken, until the rotational speed continued to decline
and no steady reading could be observed.

5. EXPERIMENTAL RESULTS

The rotor was observed to be self-starting under no load in a wind
speed of 3·5 m/s, provided the pitch variation was greater than ± 4°.
Typical data from test runs are illustrated in Fig. 11. These clearly
demonstrate a degrading of the power characteristics with an increase of
pitch variation.

During tests there was always a considerable offset to the vane at
the centre of the rotor. At times this effect was as much as 80° to the
direction of flow through the tunnel.

The rotor would accelerate to maximum speed faster with a pitch
variation of ± 15° than it did with a pitch variation of ± 5°. The
acceleration to maximum speed was in two different stages. The rotor,
at a setting of ± 5° pitch variation and with a tunnel velocity of 7 m/s,
accelerated slowly up to a speed of 120 r.p.m. (this could take 1·5 -
2 mins) then the rotor would accelerate very fast (10 - 15 secs) to its
final speed of 300 r.p.m.

At rotational speeds of 300 r.p.m. and above the rotor experienced
vibrations of the wind vane and the shaft extension.

Before test runs it was difficult to adjust the rigging so that
each wire was under the same tension and, in the case of the upper wires,
to maintain a straight shaft extension, in addition to maintaining a
straight blade.

The dynamometer in use worked well and although spring-balances were originally going to be used to apply the load, from experience these were found to be too insensitive to be consistent. Weights were used subsequently.

6. RESULTS/COMPARISON WITH THEORY

The first requirement in this comparison is to evaluate the difference between the theory and experiment. There are two main factors in the experiment that were not allowed for in the theory. Firstly there is aerodynamic/parasitic drag of the rotor, and secondly the friction of the rotor bearings and the pitch control mechanism.

7. AERODYNAMIC DRAG

Consider an element of length dr at a radius r from the centre of the rotor, along a support arm. The drag on this element is

$$dD = \tfrac{1}{2} \, \rho \, C_D \, c \, (wr)^2 \, dr$$

hence the torque induced by this drag is

$$dT_D = \tfrac{1}{2} \, \rho \, C_D \, c \, w^2 \, r^3 \, dr$$

Integrating from r = 0 to R gives the drag torque caused by one support arm.

$$T_D = \tfrac{1}{8} \, \rho \, C_D \, c \, w^2 \, R^4 \qquad (3)$$

A similar expression can be obtained for the rigging wires.

The drag torque associated with blade fittings, particularly the junction at the blade root, can also be shown to be proportional to w^2.

Total drag torque of rotor $= K_A N \rho \, w^2 \, R^4 \, c$ where K_A is a function of the geometry and drag characteristics of the various components of the rotor (e.g. wires, connections, support arms, etc.).

Thus total power losses due to drag becomes

$$K_A N \rho \, w^3 \, R^4 \, c \qquad (4)$$

Thus loss in power coefficients due to aerodynamic drag

$$\Delta \, C_{P_{aero}} = K_A \frac{N \, X^3}{A} \qquad (5)$$

As a result of the first set of tests in the wind tunnel at Bath, it was felt that considerable power losses were coming from the aerodynamic drag around the blade root area including the control rod (Fig. 12). This was confirmed by tests on a full scale model in the 18 in x 18 in working section of a wind tunnel at University of Exeter (Fig. 13). The tests on these models were performed without any movement of the blades and so could only be comparative. However, it was estimated that the power coefficient of the rotor could be improved by up to 50% by the improvement of the aerodynamics of this region and by putting the control rod inside the blade support arm. Fig. 14 shows a vertical velocity profile plot in the region behind the blade axis. On the left hand side is that caused by the original blade root shape, and on the right is the profile caused by the later configuration.

Further aerodynamic improvements made include the replacement of 2 mm dia. wire rigging by lenticular rod rigging; and the addition of aerodynamic blisters to shroud the connection of the rigging to the blades.

8.   FRICTION

Frictional losses at the blade root bearing are a function of the thrust and radial loads on that bearing.

Radial load = $h_1 \times \rho_{al} H c^2 R w^2$

where h_1 and h_2 are a function of the blade internal geometry and the blade support locations.

Thrust load = $h_2 \rho_{al} H c^2 R w^2$

$h = h_1 + h_2$

Frictional Torque due to these loads is

$$T_F = h \mu \rho_{al} H c^2 R w^2 \bar{r} \tag{6}$$

where $\bar{r}$ is the mean radius at which friction acts

Blade position relative to tangent of blade path

$$\gamma = a \sin wt$$

$$\therefore \quad \dot{\gamma} = a w \cos wt$$

Power loss due to friction = $T_F \times \dot{\gamma}$

$$\text{Average Power loss} = \frac{1}{\pi/2w} \int_{0}^{\pi/2w} T_F \, a \, w \cos wt \, dt$$

$$= \frac{2 a w}{\pi} \cdot T_F$$

$$= \frac{2h \, a \, \mu \, \rho_{al} \, H c^2 \, R \, w^3 \, \bar{r}}{\pi} \tag{7}$$

If $\bar{r}$ is a function of c, loss in power coefficient due to friction is:

$$\Delta C_{p_{fric}} = K_F \, a \, \frac{S^3 \, X^3}{N^2} \tag{8}$$

The offset of the vane is an indicator of bearing friction and an unwise choice of bearing material (molybdenum disulphide impregnated nylon) together with a more complex mechanism as a result of putting the control rod of the blade inside the support arm gave a large element of frictional resistance to the rotor. In addition, if the blade and blade movement system is not very carefully balanced, large forces will be transmitted along the control arm to the eccentric bearing.

9. COMBINED POWER LOSSES

From equations (5) and (8) it is now possible to write down a formula expressing the power coefficient losses that should correlate the experiments with the theory.

$$\Delta C_{p_{losses}} = (K_A \frac{N}{A} + K_F \frac{a\ S^3}{N^2})\ X^3 \qquad (9)$$

In Fig. 15 the difference between the theoretical and the experimental results are plotted against the tipspeed ratio. Least squares cubic curves have been fitted to the sets of data shown. Examination of the curves together with the formula for power coefficient losses indicate that aerodynamic drag accounts for 30% and friction accounts for the remainder of the total lost.

At a tipspeed ratio of 3.5 the experimental power losses diverge from the least squares curve for tipspeed ratios lower than that mentioned. This effect corresponds to a power coefficient of 20%. Consideration of this effect, together with aerodynamic improvements of the rotor that should have improved the power coefficient in experiments suggest that there is an effect in the wind tunnel that prevents the rotor from extracting any more than 20% of the power through the rotor. It is believed that the wake of the rotor is unable to expand any further than the size corresponding to a 20% power extraction. Evaluation of the size of the wake and its deflection as it passes through the rotor and an allowance for the boundary layer along the walls of the wind tunnel are illustrated in Fig. 16.

In spite of this major problem to the work in the wind tunnel the results are useful, particularly when the overall power coefficient is less than 20%.

10. CONCLUSIONS

A theoretical model of the performance of the variable pitch vertical axis wind turbine is operating on the computer and these results indicate the characteristics of the machine. Experimentally there are losses that have to be allowed for and the reduction of these losses suggest that maximum power coefficients of over 30% are obtainable with this machine. Our experience in the wind tunnel illustrates a considerable drawback in closed flow tunnels for wind turbine power evaluation and suggests that blockage does not give an increased speed at the rotor, but rather higher speed in the wake than anticipated. We are planning to test the rotor unaltered on a lorry test bed to examine its performance in unconfined air, to confirm the existence of blockage in the wind tunnel. Using results from these tests we intend to construct a mark II machine with greatly reduced friction characteristics, as well as reduced aerodynamic drag.

11. ACKNOWLEDGEMENTS

The authors would like to thank Professor Black and his staff at the School of Engineering, University of Bath for the use of the wind tunnel, and also for much useful advice. Thanks are also due to Professor Lacey of the Department of Chemical Engineering, University of Exeter, without whose support this work would not have been done. Particular thanks are due to Derek Mayes, who constructed the rotor in the Faculty Workshops, and to the secretarial staff of the Department of Chemical Engineering for typing.

12. <u>REFERENCES</u>

1. BRULLE, R.V. and LARSEN, H.C.   Giromill, Investigation for generation of electrical power. Wind Workshop 2, Washington, June, 1975.

2. STRICKLAND, J.H.   A performance prediction model for the Darrieus turbine.   Paper C3, International Symposium on Wind Energy Systems, Sept. 1976, Cambridge.

3. SHARPE, D.J.   An Aerodynamic Performance Theory for the Darriens Wind Turbine.   Paper S4, International Symposium on Wind Energy Systems, Sept. 1976, Cambridge.

4. SHELDAHL, R.E. and BLACKWELL, B.F.   Aerodynamic Characteristics of four symmetrical airfoil sections through 180 degree angle of attack at low Reynolds Numbers.   Paper at Vertical Axis Wind Turbine Workshop, Albuquerque, N.M., 1976.

5. SARRE, P.E.   Internal Report No. WEP1, Wind Energy Project, March 1977.

Table 1
Parameters specified for computer run

Rotor radius   Number of blades

Chord length   Rotor height

Maximum pitch   Position of maximum pitch

Average pitch   Aerofoil data to be used

Incorporation of drag characteristics

Range of tipspeed ratio through which calculations are performed

Density of calculations

Output option

Table 2
Rotor specification

2.4m diameter X 1.6m high variable pitch vertical axis wind

turbine

Aerofoil: NACA 0015   Chord = 0.145m

Pitch variation 0°- 20° [to be preset]

Average pitch   0° -30° [to be preset]

Pitch actuated by eccentric bearing

Lenticular rod rigging

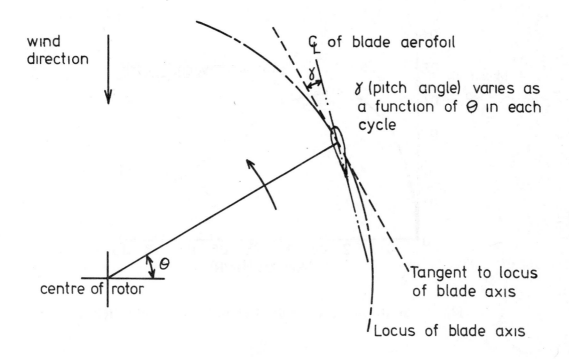

wind
direction

Ç of blade aerofoil

γ (pitch angle) varies as
a function of Θ in each
cycle

Tangent to locus
of blade axis

Locus of blade axis

centre of rotor

Θ

Fig. 1   Description of blade position.

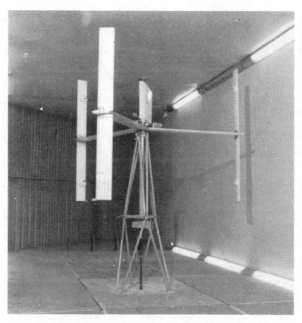

Fig. 2   2.4m diameter x 1.6m high variable pitch
vertical axis wind turbine in the wind tunnel.

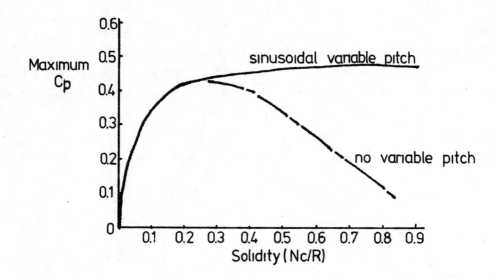

Fig. 3 Maximum power coefficient as a function of solidity.

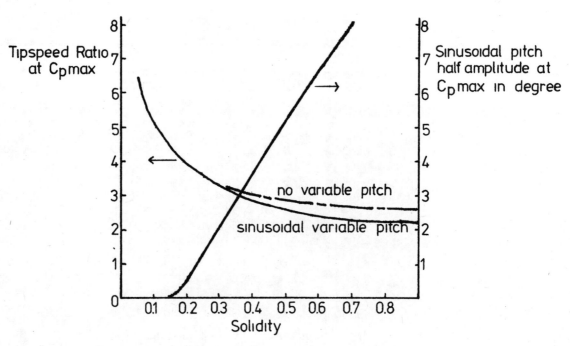

Fig. 4 Tipspeed ratio and sinusoidal pitch half
amplitude at C max as a function of solidity.

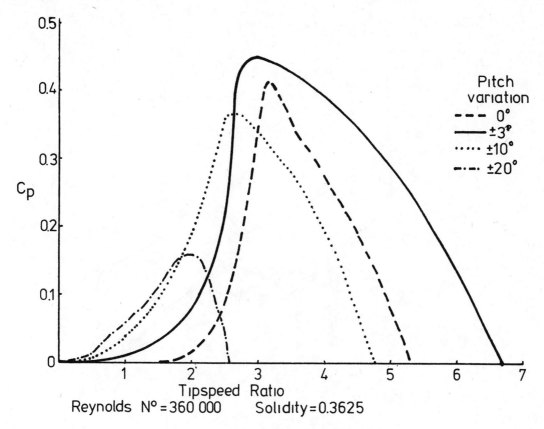

Reynolds N° = 360 000   Solidity = 0.3625

Fig. 5 C_p versus tipspeed ratio for
a variety of pitch half amplitudes.

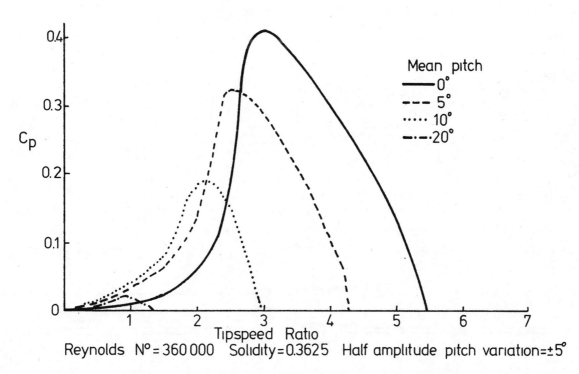

Reynolds N° = 360 000   Solidity = 0.3625   Half amplitude pitch variation = ±5°

Fig. 6   Cp vs Tipspeed ratio for a range of mean pitch.

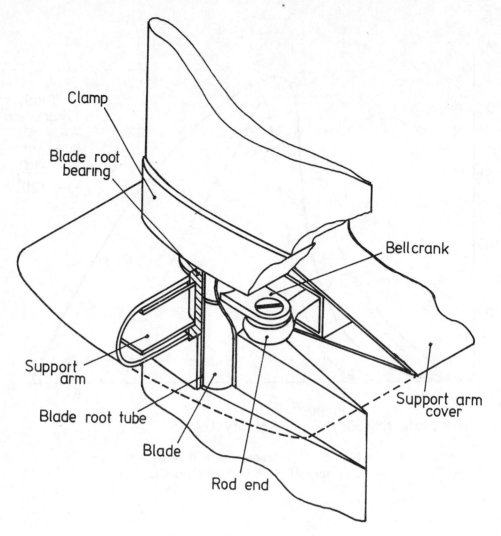

Fig. 7 Blade root detail.

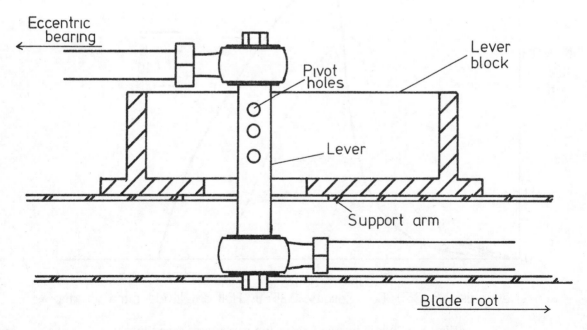

Fig. 8 Drawing of lever system.

Fig. 9   Photograph of dynamometer.

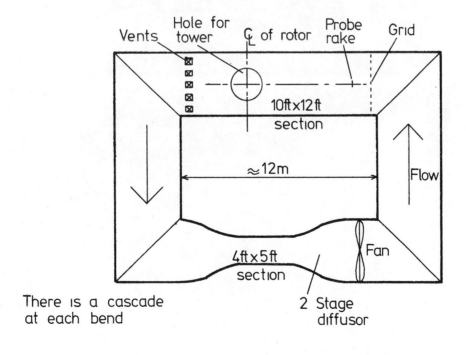

Fig. 10   Layout of wind-tunnel.

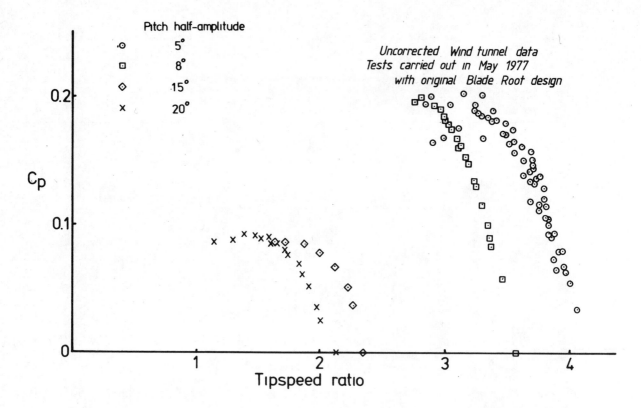

Fig. 11   Test data from wind-tunnel on 2.4 m diameter x 1.6 m VAWT.

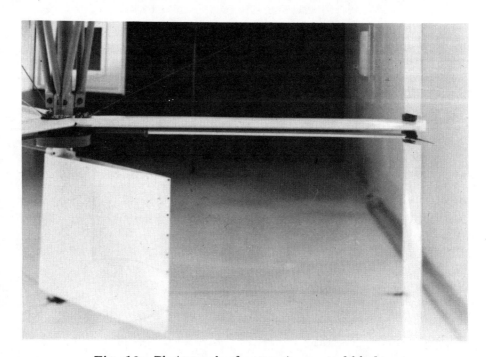

Fig. 12   Photograph of support arm and blade.

Original blade root   Later low drag blade root

Fig. 13   Blade root models tested in 18 ins square wind tunnel.

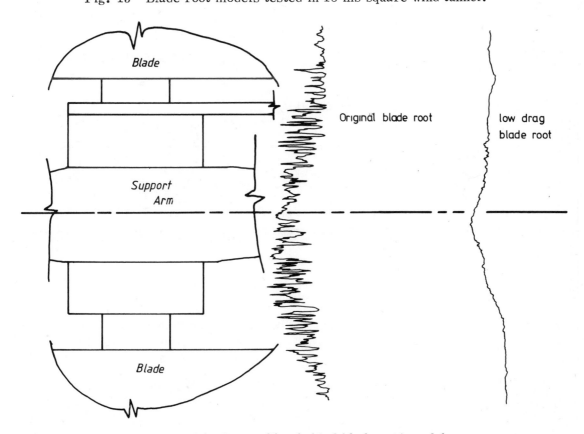

Fig. 14   Velocity profiles behind blade root models.

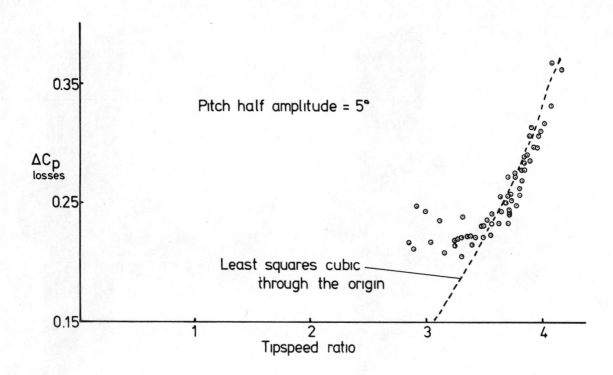

Fig. 15   Plot of experimental power coefficient losses vs tipspeed ratio.

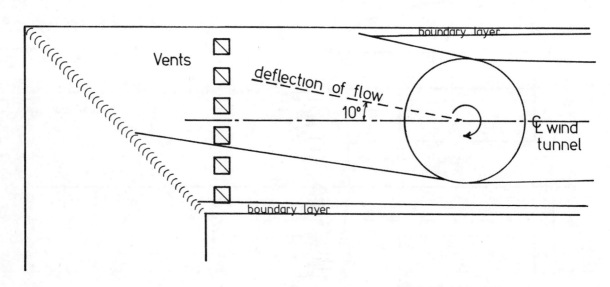

Fig. 16   Wind tunnel walls and boundary layer superimposed
on free air flow around VAWT extracting 20% power from the wind.

WIND ENERGY SYSTEMS

October 3rd - 6th, 1978

THE AERODYNAMIC PERFORMANCE OF THE SAVONIUS ROTOR

G.M.Alder

University of Edinburgh, U.K.

Summary

The paper describes wind tunnel tests on a Savonius rotor. The rotor was of high aspect ratio with blades of semicircular cross-section. The mean and the periodic components of torque, drag force and side force were measured. No significant differences in dimensionless performance were observed for Reynolds numbers (based on wind speed and rotor swept diameter) in the range 2.10^5 to 6.10^5. The mean drag coefficient, and the amplitudes of the periodic drag and side-force coefficients took values of about 1.0 over most of the rotation speed range. A significant alteration in the nature of the flow round the rotor occured at a tip-speed ratio of 0.2, where the mean side-force changed sign and changes took place to the shapes of the periodic wave forms of all the performance parameters.

Held at the Royal Tropical Institute, Amsterdam, Netherlands.

Symposium organised by BHRA Fluid Engineering in conjunction with the Netherlands Energy Research Foundation ECN

©BHRA Fluid Engineering, Cranfield, Bedford, MK43 0AJ, England.

NOMENCLATURE

b   - blade overlap

c   - blade chord

C_q - torque coefficient

C_x - drag coefficient

C_y - side-force coefficient

D   - rotor swept diameter

H   - height of wind tunnel test section

L   - width of wind tunnel test section

p_o - gauge pressure in tunnel settling section

Q   - torque

V   - wind velocity

X   - drag force

Y   - side force

β - tunnel blockage ratio (D/H)

η - rotor efficiency

θ - inclination of rotor to wind

λ - tip-speed ratio

ν - kinematic viscosity of air

ρ - density of air

ω - angular velocity of rotation

INTRODUCTION

Interest in vertical axis wind turbines has developed considerably during the last few years. Attention has centred particularly on the Darrieus rotor because of its good aerodynamic efficiency and because its high rotation speeds are appropriate to the generation of electric power. This type of machine is also amenable to theoretical modelling. Consequently, a considerable body of theoretical and experimental data exist on Darrieus turbine performance.

The Savonius rotor is widely used in small-scale applications where the advantages of cheap and simple construction, and high starting torque outweigh its lower efficiency. It has also been proposed as an auxiliary device for starting a Darrieus turbine (Ref. 1) and as a tidal-power generator (Ref. 2). Nevertheless, the picture presented in the literature of Savonius rotor aerodynamics and performance (eg. Refs. 2, 3 & 4) is far from complete. The published studies of rotor geometry (Refs. 2 & 4) show that the optimum Savonius rotor has two semicircular blades overlapping by about a quarter of their chord. The optimum power efficiency (expressed as a fraction of the Betz limit) is then about 0.2, and the rotor drag coefficient is about 1.0. There is very little information available concerning the more general aspects of the flowfield round a Savonius rotor; for example, the role of the slot between the blades is not clear. A flow-visualisation study is discussed in Ref. 2, but it is inconclusive. The system does not appear to lend itself to theoretical analysis, a factor which greatly inhibits the search for the most satisfactory rotor geometry.

This paper describes the first stage of an experimental investigation into the aerodynamics of a Savonius rotor. A two-dimensional (ie. high aspect ratio) rotor with two semicircular blades was tested in a wind tunnel at Reynolds numbers (based on wind velocity and rotor swept diameter) of about 4.10^5. The parameters studied were the torque, the drag force (in the wind direction) and the side force (perpendicular to the wind). All three parameters have mean and periodic components which are functions of rotation speed. The results were corrected for tunnel blockage.

APPARATUS

The experiments were carried out in an open-circuit wind tunnel with a closed test section which was 0.9 m long and 0.6 m x 0.3 m in cross-section. The settling section of the tunnel was 0.9 m square and 4 m long, and was supplied with air by a centrifugal fan powered by a 50 kW variable speed DC motor. Test section velocities used here lay in the range 8-25 m/s. The velocity profile in the empty test section was flat to within ± 2% and the turbulence intensity (r.m.s. fluctuations/mean velocity) was about 0.4%.

The rotor model was 0.3 m long and spanned the centre of the test section. It consisted of blades rolled from light alloy sheet and attached to circular end plates which were flush with the tunnel wall. The air velocity approaching the rotor was measured by a pitot-static tube on the tunnel centre-line 0.45 m upstream from the rotor axis. Fig. 1 shows the general arrangement of the rotor and its associated instrumentation. The rotor was carried by self-aligning bearings, one of which was mounted on a two-component force balance and the other at the centre of a flywheel. The flywheel was carried by a separate shaft on rigid mountings. The torque between rotor and flywheel was transmitted through a torque meter whose output voltage was then amplified before passing through slip-ring contacts to the recording instruments. Moments of inertia about the rotation axis were 0.012 kg.m^2 for the rotor and 1.02 kg.m^2 for the flywheel. The flywheel shaft was also coupled to a permanent-magnet motor (Printed Motor type GPM 12M4-B) which could either drive or load the shaft so that the rotation speed of the whole system was independent of the aerodynamic torque generated by the Savonius rotor.

The drag and side forces, and the torque were measured by strain-gauged balances of the double-cantilever type. The lowest resonant frequency of the rotor and measurement system was 53 Hz where the vibration mode was mainly rotation about the axis. This figure set an upper limit to the rotation speed at which the periodic torques and forces could be measured. With a two-bladed rotor, interest lies in the even harmonics of rotation speed. The 2nd, 4th and 6th harmonics are probably the most significant, and to obtain these with reasonable accuracy, the periodic measurements were taken at rotation speeds less than 5 Hz. The DC components of the torque and force measurements were passed directly to a pen recorder, together with a speed measurement signal. The AC components were first stored in a digital oscilloscope and then transferred to the recorder.

The rotor blades were rolled from 2 mm light alloy sheet and soldered to 2 mm flanges, which in turn were bolted to the end-plates carrying the bearings. The cross-section of the rotor is shown in Fig. 2. The blades were semicircular, with a chord (c) of 135 mm and an overlap (b) of 41 mm. This degree of overlap has been found (Refs. 2 & 4) to be optimum for this type of rotor.

RESULTS AND DISCUSSION

The force and torque results may be expressed as dimensionless coefficients defined as follows:

$$\text{drag:} \qquad C_x = \frac{X}{\frac{1}{2}\rho DLV^2}$$

$$\text{side force:} \qquad C_y = \frac{Y}{\frac{1}{2}\rho DLV^2}$$

$$\text{torque:} \qquad C_q = \frac{Q}{\rho LD^2 V^2}$$

For a rotor of given geometry, these coefficients will in general be functions of the tip speed ratio $\lambda (= \omega D/2V)$ and the Reynolds number (VD/ν). The rotor efficiency, expressing power as a fraction of the Betz limit, is given by:

$$\eta = \lambda C_q /0.593$$

The relative geometries of the rotor and the wind tunnel test section are also important, the most significant parameter being $\beta(= D/H)$, the ratio of the rotor swept diameter to the tunnel height. The present work is also concerned with the periodic components of forces and torque which depend on the inclination of the rotor to the wind direction, θ (defined in Fig. 2).

A preliminary investigation into the effects of Reynolds number was carried out by testing the rotor at tunnel speeds ranging from 5-25 m/s. No significant effects on the dimensionless results were observed in this range, which corresponds to Reynolds numbers of $2.10^5 - 6.10^5$.

Tunnel interference effects were important in this work because a considerable proportion of the wind tunnel cross-section was blocked by the rotor. The high value of β (0.39) was chosen in order to keep rotation speeds down and thus limit the influence of resonance in the force measurement systems. The effect of tunnel blockage was studied by constructing two additional rotors of different swept diameters giving $\beta = 0.21$ and 0.58 respectively. All three rotors were of similar cross-section geometry, with the exception of the thickness of the metal sheet from which they were rolled (1.5 mm for the smallest rotor; 2 mm for the larger two). It was found that the mean torque and drag coefficients lay within 5% for the three rotors if the wind velocity, V, was calculated directly from the pressure in the tunnel settling section (where the velocity was negligible), ie.

$$V = \sqrt{2p_o/\rho}$$

This is the air velocity in the jet at the end of the test section where it discharges into the atmosphere. Further investigation is needed to establish whether such an approach to the blockage problem is generally applicable. The effect of blockage on the periodic components was not studied. More recently, the work of Alexander (Ref. 5) has become available. It is certainly more satisfactory to base an analysis of tunnel blockage on the conditions upstream of the rotor where they are not affected by the confused flow in the wake. The drag coefficients (at $\lambda = 0.5$) for the three rotors discussed above were evaluated on the basis of upstream velocity measurements. The results are plotted against β in Fig. 3, where the corresponding values from Ref. 5 are also shown. The difference between the two sets of results can probably be explained by the fact that Ref. 5 dealt with three-dimensional flow round rotors of finite height, whereas the present work was concerned with two-dimensional flow.

The results of the Savonius rotor tests at $\beta = 0.39$ will now be discussed. Fig. 4 shows the mean (DC) components of forces and torque plotted as functions of tip-speed ratio. The efficiency is also shown in Fig. 4, and the maximum value $\eta = 0.21$ at $\lambda = 0.82$ is in general agreement with previous work (Refs. 2, 3 & 4). Fig. 5 shows the periodic (AC) components of the force and torque coefficients expressed as the peak-to-peak amplitudes. The periodic components are shown as functions of rotor inclination (θ) in Figs. 6, 7 & 8 for $\lambda = 0.05$, 0.2 and 0.4. The shapes and amplitudes of the periodic waveforms did not change significantly above $\lambda = 0.4$.

An examination of the results shows that a significant change in the rotor flowfield occurs at $\lambda = 0.2$. This is shown particularly by the side-force, whose periodic amplitude is very large at that speed, and whose wave-form changes shape, mainly by undergoing a phase-shift in the 4th harmonic of rotation speed. It is also apparent that the steady increase in mean side-force due to the Magnus effect does not occur until $\lambda > 0.2$. The particular significance of $\lambda = 0.2$ is also shown by the torque, whose mean component experiences a distinct discontinuity in the rate of change, $dC_q/d\lambda$, and whose wave-form also undergoes changes of amplitude and shape. The reasons for these events at $\lambda = 0.2$ are not clear. It is possible that they are associated with changes in the flow through the slot between the blades. This, however, does not provide a complete explanation, because testing the rotor with the slot temporarily blocked led to a 50% reduction in mean torque coefficient but to an otherwise qualitatively similar set of results.

CONCLUSIONS

Wind tunnel tests of a Savonius rotor with semicircular blades of high aspect ratio showed that the optimum efficiency occured at a tip-speed ratio of 0.82. The mean side-force and drag coefficients at this speed were 1.25 and 0.95. The peak-to-peak amplitudes of the periodic components of torque, side-force and drag coefficients were of the same order of magnitude as the mean values. Reynolds numbers in the range $2.10^5 - 6.10^5$ were found to significantly affect the results. A significant change in the nature of the flow round the rotor was found to occur at a tip-speed ratio of 0.2.

Further work on Savonius rotor aerodynamics is progressing. Attention is being directed towards obtaining a full harmonic analysis of the periodic wave-forms, and towards a study of the effects of changes in rotor geometry.

REFERENCES

1. Templin, R.J. and South, P. "Some design aspects of high-speed vertical-axis
    wind turbines." 1st International Symposium on Wind Energy Systems, BHRA
    Fluid Engineering, Cambridge, England (September 7-9, 1976).

2. Manser, B.L. and Jones, C.N. "Power from wind and sea - the forgotten panemone."
    Thermo-fluids Conference : Energy - Transportation, Storage and Conversion,
    Brisbane, Australia (December 3-5, 1975). Inst. Eng. Natl. Conf. Publ. 75/9,
    Sydney.

3. Savonius, S.J. "The S-rotor and its applications." Mechanical Engineering, 53,
    5, 333-337 (May 1931).

4. Alexander, A.J. and Holownia, B.P. "Wind tunnel tests on a Savonius rotor."
    Journal of Industrial Aerodynamics (to be published).

5. Alexander, A.J. "Wind tunnel corrections for Savonius rotors." 2nd Inter-
    national Symposium on Wind Energy Systems, BHRA Fluid Engineering, Amsterdam,
    Netherlands (October 3-5, 1978).

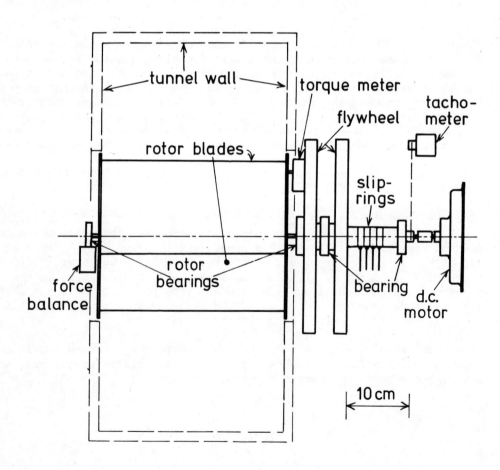

Fig. 1   Rotor and dynamometer assembly.

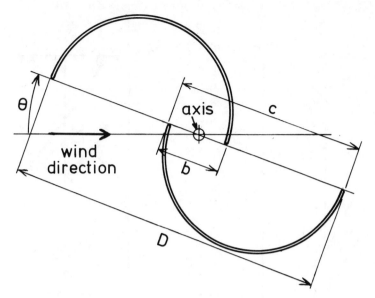

Fig. 2   Rotor cross-section geometry.

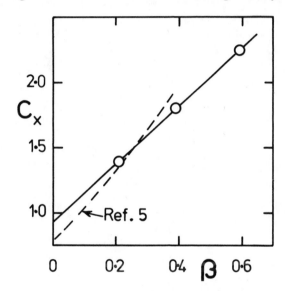

Fig. 3   Maximum drag coefficient as function of tunnel blockage ratio for three geometrically similar rotors.

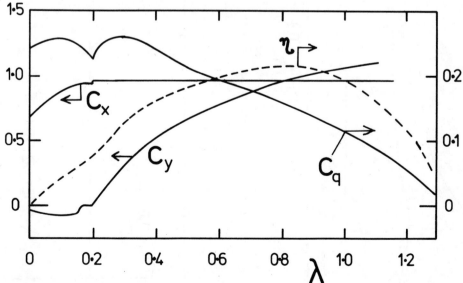

Fig. 4   Efficiency, and mean components of side-force, drag and torque coefficients as functions of tip-speed ratio.

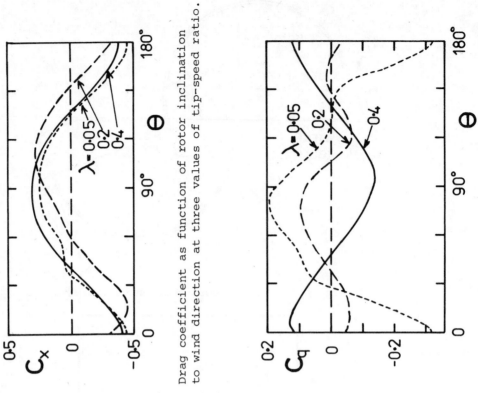

Fig. 6   Drag coefficient as function of rotor inclination
to wind direction at three values of tip-speed ratio.

Fig. 8   Torque coefficient as function of rotor inclination
to wind direction at three values of tip-speed ratio.

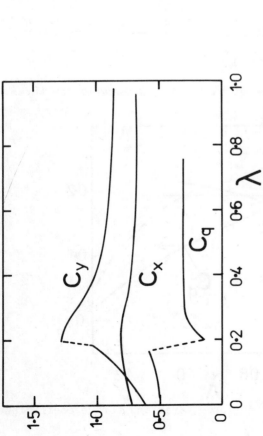

Fig. 5   Peak-to-peak amplitudes of periodic components of
side-force, drag and torque coefficients as
functions of tip-speed ratio.

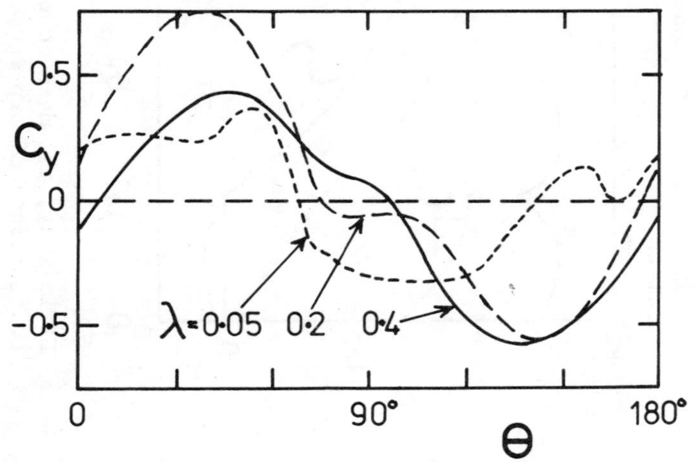

Fig. 7   Side-force coefficient as function of rotor
inclination to wind direction at three values

POWER AUGMENTATION IN A DUCTED SAVONIUS ROTOR

A. Sabzevari

Pahlavi University, Iran

Summary

Several ductings, concentrators and diffusers are examined and their effects on the performance characteristics of a split S Savonius rotor are presented in this paper.

The systematic results obtained from the performance of the several models under investigation have led to the design of a circularly ducted Savonius rotor equipped with a number of indentical wind concentrators and diffusers along the periphery of its cylindrical housing. The new design takes advantage of wind concentration, diffusion and the generated wind-draught due to the low pressure region behind the circular rotor housing cylinder. Wind tunnel results are presented over a considerable range of velocities.

Held at the Royal Tropical Institute, Amsterdam, Netherlands.

Symposium organised by BHRA Fluid Engineering in conjunction with the Netherlands Energy Research Foundation ECN

©BHRA Fluid Engineering, Cranfield, Bedford, MK43 0AJ, England.

NOMENCLATURE

A	area of the air tube
A_s	rotor swept area
c_p	power coefficient
D	rotor diameter
h	rotor height
R	rotor tip radius
r	rotor vane radius
T	output torque
V	undisturbed wind velocity
θ	tilt angle of diffuser
λ	tip speed ratio
ρ	air density
ω	rotor angular velocity

1-The power coefficient c_p employed in this work is defined as the ratio of the measured output shaft work devided by the energy content of the airtube at the inlet section of the concentrator, i.e. $T\omega/\frac{1}{2} \rho A V^3$.

2- The-tip-speed ratio λ is defined as the ratio of the tip of the vane velocity devided by the undisturbed wind velocity, i.e. $\frac{r\omega}{V}$.

INTRODUCTION

In an earlier paper by the present writer (Ref. 1) the performance characteristics of several ducted models of the Savonius rotor, Fig. (1), have been presented. It has been shown that the ducted rotors, especially when equipped with wind concentrator of the type shown in Fig (1d), result in more favourable values for both the tip-speed-ratio λ and the power coefficient c_p compared to those of free Savonius rotor. The diffuser of the type shown in Figs $(1e,f)$ did not hower tend to improve the performance of the ducted rotor model with concentrator.

Further examinations of the effect of concentrators and diffusers (Ref. 2) on the performance of split Savonius rotor have resulted in the followings:

i) While concentrators of the type (a) in Fig. 2 are quite ineffective, those of the type (b) in the same figure yield considerable improvement in the rotor performance. Since the area, A, of the incident stream tube with velocity V at inlet to the concentrator of the former type is considerably higher than the rotor swept area, then in many occasions even decreases in values of c_p (compared to those of free rotor) has been observed. This result is in conformity with some of discussions presented at the First International Conference on WES (Ref. 3). The concentrator of the latter type, Fig. 2b, however has an inlet area exactly the same as the swept area of the rotor itself. This is achieved by covering up the convex portion of the rotor vane and then concentrating mechanically the air-flow on the concave half of the rotor as shown in the figure.

ii) Among the three types of diffuser arrangements, Fig. 3, tested in the wind tunnel, only model 3c was shown to improve the Savonius rotor performance. Furthermore the diffuser tilting angle, θ, itself was an important parameter; the tilting angle of about 33° gave the best results over the range of wind velocities tested. This value of θ is in confirmity with an earlier statement made by Tokaty, (Ref. 4) concerning the flow behind the free Savonius rotor.

While the improving effects of ducting the split Savonius rotor and the installation of special concentrator and diffuser designs have been demonstrated in references 1 and 2, the important question of the natural wind direction at site with respect to the arrangements presented so far is still unresolved. That is to say, while wind may blow in any direction for the free Savonius rotor, that for the ducted arrangements shown should only be along the longitudinal direction of the concentrator.

The aim of the present work is to investigate some further ducting arrangements for the split Savonius rotor with several important qualities in mind. These basic qualities are to avoid dead air regions as exist in the corners of the rectangular ducts, and while maintaining the advantages of guided wind concentration and diffusion the design of the rotor housing should be fairly independent of wind flow direction.

To achieve the above-mentioned aims, the split S rotor is proposed to be encircled by a circular cylinder with a number of identical concentrators and diffusers placed along the periphery of the cylinder as suggested in Fig. 4.

CIRCULARLY DUCTED MODELS

In order to be able to separate various effects on the overall performance characteristics of the proposed ducted rotors of Fig. 4 seven models, as presented in Fig. 5, are tested in the subsonic wind tunnel. The split S rotor in all seven models is the same with dimensions of; D = 32 cm, h = 40 cm and r = 10 cm. The terms D, h and r denote the rotor diameter, rotor height and the radius of each half-circular cylinder respectively. The opening width of each concentrator or even diffuser is 32 cm with the height exactly the same as that of the rotor itself, thus the area of undisturbed flow tube before being mechanically concentrated is the same as the rotor swept area. All seven models of Fig. 5 are placed between end plates.

The performance of model 5a forms the base for comparison. While model 5b demonstrates the effect of flow concentration, models 5c and 5d will show the influence of diffusion in right and left diffuser respectively. Models 5e and 5f have respectively 2 and 3 identical openings in each of which, depending on the wind direction, one opening acts as concentrator and the other one or two as diffusers. In model 5g the concentrators and diffusers are reproduced by 3 static columns in such a way that the outside arcs of the columns form also another broken circle. A general flow pattern over model 5g is shown in Fig. 6 in which the points of flow separation and the region of low pressure behind the outer cylinder are marked. The relatively low pressure wake of the flow over the broken cylinder would cause greater pressure difference between the concentrator inlet and the diffuser outlets with the consequence of a further drive for the rotor. This effect would be demonstrated by a comparison between the performance characteristics of models 5f and 5g.

In these ducted models the possibility of dead air formation, as existed in the rectangular ducts, is basically eliminated. Further-more the bahaviour of the model 5g is identical for 3 different wind directions as it has 3 identical openings. The static columns in windmills of this type or the type with 5 openings may be constructed from any available material at site and some artistic care could give quite a pleasing appearance to the windmill.

WIND TUNNEL TEST RESULTS

The seven models presented in Fig. 5 were tested in a subsonic wind tunnel over a wide range of wind velocities, measuring the rotor output work in terms of its torque T and its rotational speed ω for any tip-speed - ratio $\lambda = (R\omega/V)$. The coefficient of performance was then calculated as the ratio of the measured shaft work devided by the energy content of the air tube at the inlet section of the concentrator, i.e. $T\omega/\frac{1}{2}\rho AV^3$, where A denotes the concentrator inlet area which has been designed to be the same as the rotor swept area A_s.

Some of the model performances in the wind tunnel are plotted in the form of $c_p - \lambda$ curves and are presented in Figs. 7 through 10. Fig. 7 showes the effect of circular duct and wind concentrator (model 5b) on power augmentation of the Savonius rotor. The power coefficient curve of the free Savonius rotor is also added to this figure to form a base for comparison. It should be also added that the performance of model 5b is superiour to that of the similar model with a rectangular duct arrangement.

Among the two diffuser arrangements, only model 5d had some positive influence in increasing the rotor coefficient of performance, c_p. Such a result is of course due to diffuser orientation.

Figs. 8 and 9 give the c_p-λ curve for models 5e and 5f respectively. The values of c_p in Fig. 8 are some what lower than the corresponding values in Fig. 7. This is due to the fact that in model 5e when one of the openings acts as a concentrator the orientation of the exit opening is not quite correct for the positive wind diffusing influence. However, the design of the model is such that two directions are allowed for the wind. And, for model 5f this is extended to 3 directions, with the rotor performance of Fig. 9.

Finally, the variation of power coefficient with the tip-speed-ratio for model 5g is plotted in Fig. 10 on which the corresponding c_p values for the free Savonius rotor and model 5f are also superimposed. As it is observed the low pressure region of the flow wake has positively added to the amount of power output. It must be added however that for the c_p values presented in Fig. 10, the value for area A has been taken as before.

The preliminary test results indicate similar behaviour for circular duct design with 5 openings as proposed in Fig. 4b.

CONCLUSIONS

From the experimental investigations carried out in this work with a number of ducted Savonius rotor models, the following conclusions have been drawn:

1. Circular ducting, wind concentrator of type model 5b and wind diffusser of the form Fig. 5d have in turn significant improvement on the performance of the split S rotor.

2. Incorporation of the concentrator and diffuser of types 5b and 5d in a circularly ducted rotor yields most favourable values for c_p, but in such a design a yawing motion device is required to bring the windmill always into correct position with respect to natural wind direction.

3. The proposed windmill models of Fig. 4, the design of which takes advantages of wind concentration in front and wind difffusion in the rotor back as well as further rotor drive, due to flow separations from static columns and low pressure wake down stream, have shown encouraging performances.

ACKNOWLEDGEMENTS

The author is thanksful to H. Moghimi and A. Ketani, students of mechanical engineering department in helping with the model tests. The author acknowledges with gratitude the useful discussion he has had with Professor G.A. Tokaty on the Savonius rotor aerodynamics. Fine typing effort of Mrs. Shatterpouri is also acknowledged.

REFERENCES

1. Sabzevari, A.: "Performance characteristics of concentrator power augmented Savonius windmill", Wind Engineering, 1, 3,..., 1977.

2. Sabzevari, A.: "Effects of concentrators and diffusers on the performance of split S rotor", Research Report, Department of Mechanical Engineering, Pahlavi University, 1977.

3. Van Holten, Th.: "Windmills with diffuser effect induced by small tipvanes", International Symposium on Wind Energy Systems, paper E3, Organized by BHRA, Cambridge (September 7th - 9th. 1967).

4.
4. G.A. Tokaty: A History and Philosophy of Fluidmechanics, G.T. Foulis & Co. Ltd, 1971.

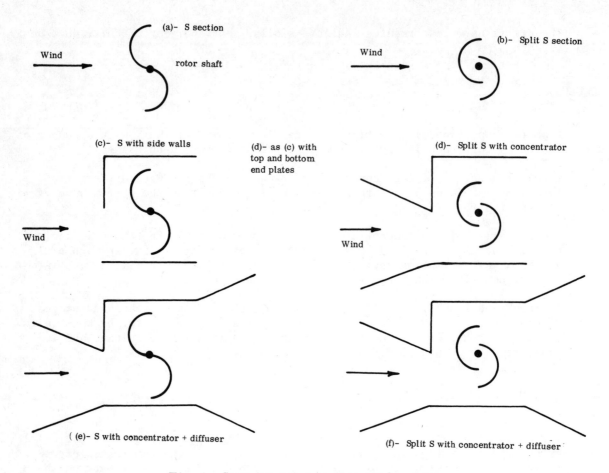

Fig. 1　Savonius rotor in rectangular ducts

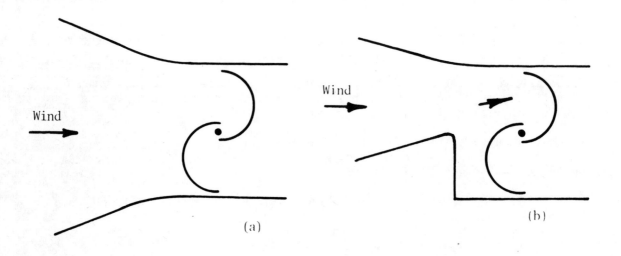

Fig. 2　Two types of concentrators

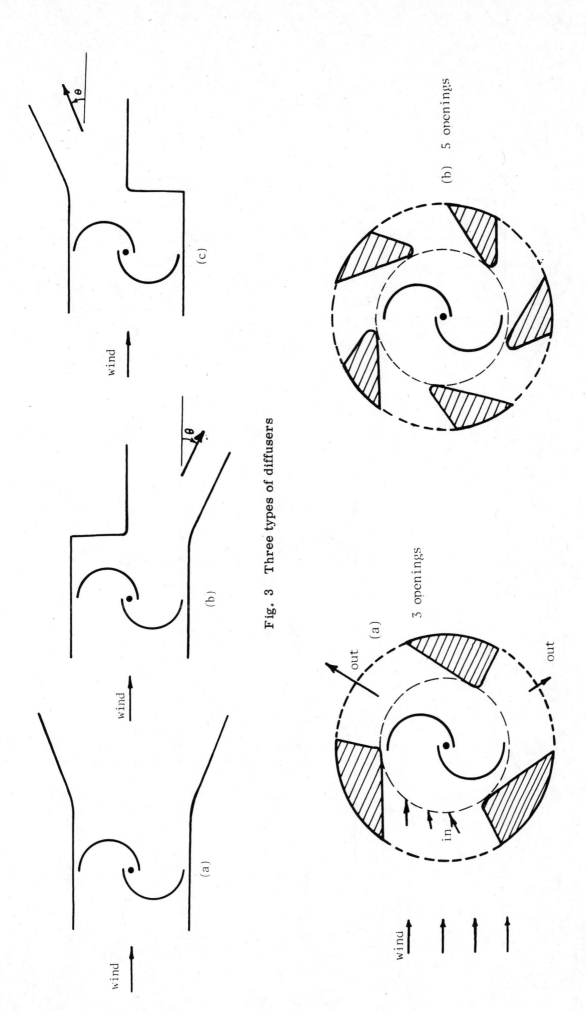

Fig. 3 Three types of diffusers

(a)

(b)

(c)

wind

wind

wind

(a) 3 openings

(b) 5 openings

Fig. 4 Proposed windmill configurations

wind

out

out

in

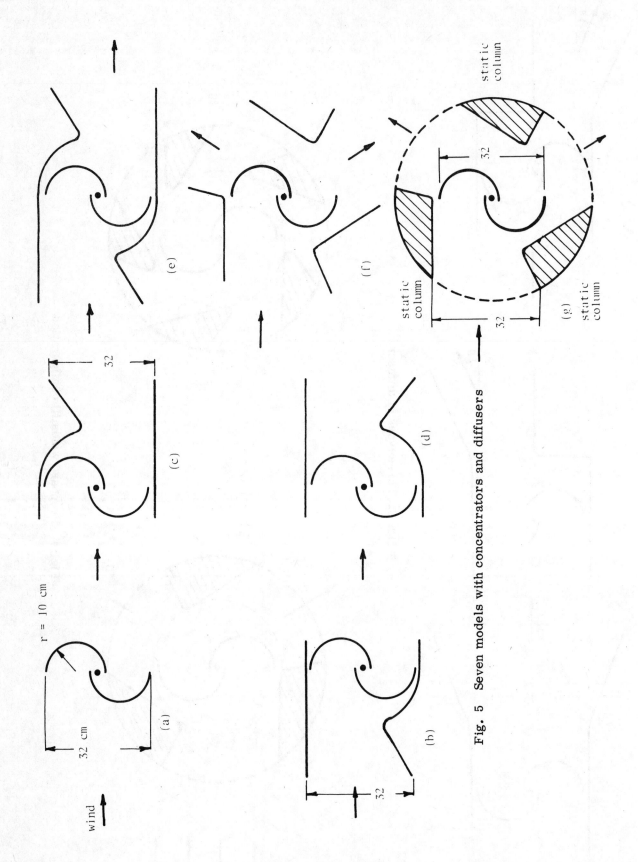

Fig. 5 Seven models with concentrators and diffusers

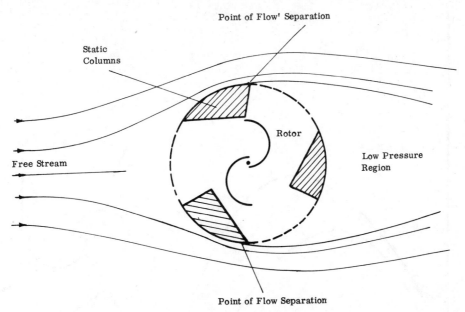

Fig. 6 A general flow pattern

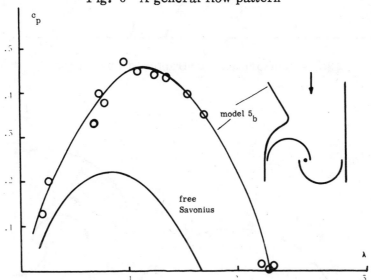

Fig. 7 The effect of concentrator on rotor performance characteristics

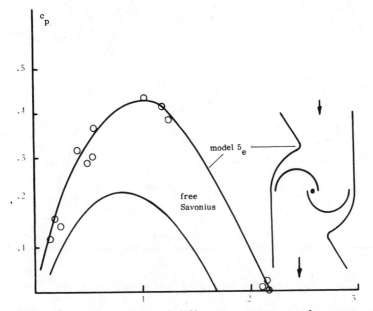

Fig. 8 The effect of concentrator and diffusor on rotor performance characteristics

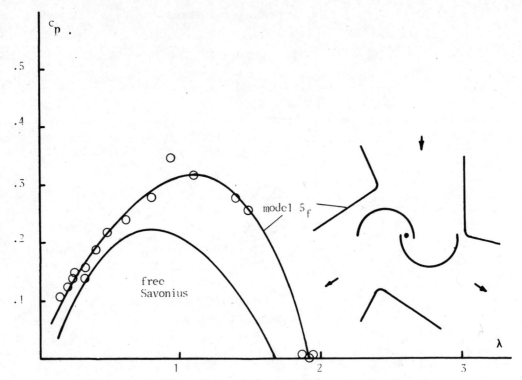

Fig. 9 Performance characteristics of model 5f

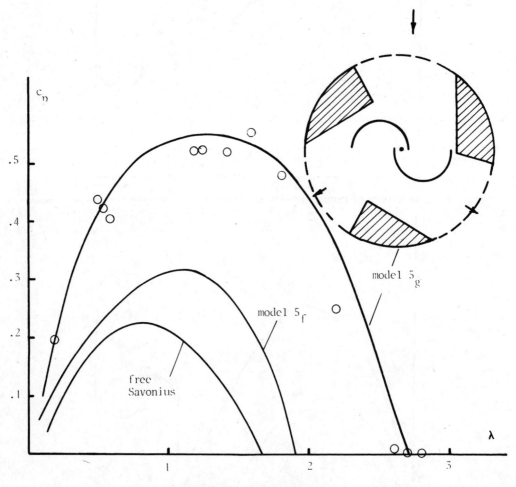

Fig. 10 The effect of low pressure wake

Discussion & Contributions

SESSION A : NATIONAL PROGRAMMES

Chairman: H.S.Stephens, BHRA Fluid Engineering, U.K.

Papers:

A1   The Netherlands research program on wind energy
             G.G.Piepers and P.F.Sens, Netherlands Energy Research
             Foundation ECN

A2   The Swedish wind energy programme
             S.Hugosson, National Swedish Board for Energy Source
             Development (NE)

A3   Recent developments in wind energy
             L.V.Divone, Department of Energy, U.S.A.

A4   The development of wind power plants in Japan
             I.Ushiyama, Ashikaga Institute of Technology, Japan

A5   Windpower programmes in Tanzania
             R.S.Reichel, University of Dar Es Salaam, Tanzania

      Notes:

             The papers were presented by the Authors whose names
             appear in bold print.

DISCUSSION

GENERAL

The following questions were directed at the authors of papers A1, A2 and A3. The authors' replies are reproduced in the appropriate sections.

N.H. LIPMAN, READING UNIVERSITY, U.K.

Are we being unkind to wind energy in applying conventional wisdom? We ask for equality in cost and a hard energy supply. Should we not also consider in an unstable world the advantage of a safe (if fluctuating) energy supply?

M. SERT, TUBITAK–MARMARA RESEARCH INSTITUTE, TURKEY

I would like to inquire in a little bit more detail about the proposed experimental schemes, particularly about the question of storage. Given the variability of the Wind Power, energy must be stored by pumping water or compressing air, etc. in order to smooth out the energy output, or, detailed systems analysis studies must be carried out to balance the output from the entire system comprising many WECS, and, I think the choice will affect the cost to a considerable extent. Mr. Divone mentioned the possibility of using wind power only for peak demand periods without any need for storage. I think, in that case, one still needs sophisticated central control mechanisms to smooth out the variations of power, frequency etc. which would require detailed consideration in cost analysis. I would request clarification on these points, and, on a smaller point of detail, that is, how are the experimental prototype designs we have been shown in slides, going to align themselves with the wind? – in smaller machines flaps were used, but it would appear that a feedback control mechanism connected to wind sensor would be required to rotate the huge rotors.

D. LINDLEY, TAYLOR WOODROW CONSTRUCTION LTD, U.K.

With respect to the storage issue it seems to me that mankinds greatest needs are for water, food, shelter and energy in that order. There are many large areas of the world without a utility grid where renewable resources of energy might be used for water or food production (e.g. fresh water from sea water by osmosis, greenhouse production of food crops etc) as well as energy production.

Wind energy conversion devices are of course capable of producing power or energy which can be translated into any of the aforementioned end uses. If, power is required storage will almost certainly be required; would Mr. Divone and the rest of the panel of speakers comment on the status of storage R and D in countries they represent.

PAPER A1

The Netherlands research program on wind energy

G.G.Piepers and P.F.Sens
Netherlands Energy Research Foundation ECN

Authors' reply to question by Professor N.H.LIPMAN.

The economic stability in Holland and in many other countries is to a large extent dependent on an uninterrupted supply of fuel from abroad. Any substantial stagnation of these supplies may lead to critical situations like those during the energy-crisis in 1973. One of the main advantages of the wind as energy source is that it cannot be influenced by disorders else where in the world. For this reason wind energy may be of interest even though the costs are not competitive with those of other energy sources.

M. LODGE, INSTITUTE OF MAN AND RESOURCES, CANADA

In light of the fact that Northern Europe is not self-sufficient in food supplies, have you included in your wind energy research and development the prospect of using W.E.C.S. to generate heat for direct use in greenhouses? If so, what projects do you have or plan in this area?

Authors' reply

In Holland greenhouses are mainly heated by means of cheap domestic natural gas with a rather poor efficiency. At the moment their portion of the total energy consumption is about 4%. In this field considerable fuel savings can be obtained. Since the bulk of the greenhouses is situated in the windy areas of Holland, WECS could play an important role. Within the frame-work of Netherland's Research Programme on wind energy a demonstration unit will be built and tested in which the wind will be used for direct heating of a block of greenhouses. The Ministry of Agriculture will participate in the project.

O. LJUNGSTROM, AERONAUTICAL RESEARCH INSTITUTE OF SWEDEN

On offshore systems in the Netherlands – what is the feeling since land is so scarce?

Authors' reply

Siting of WECS is a key problem in Holland due to its very dense population. For this reason a study concerning the installation of wind turbines in the North Sea is being made. Three regions within the Dutch part of the Continental Shelf seem to be favourable with respect to wind characteristics, wash of the waves, nature of soil and navigation routes. However, the cost of the foundations of the erection of the complete installation, of the cables for transport of the electricity and of the maintenance are an order of magnitude higher than in the case of land based units.

Authors' reply to question by Dr. M.SERT.

When WECS are used in a proper way in connection with convential power stations no separate storage systems are needed. It certainly requires a thorough system analysis in order to save a substantial amount of fossil fuel in this way.

The rotor of a wind turbine with a horizontal axis is aligned in the wind by means of a yawing device. The drive motor is actuated through a feedback control system by a wind direction sensor.

Authors' reply to question by Dr. D.LINDLEY

In Holland R & D work on storage is restricted to flywheel systems.

PAPER A2

The Swedish wind energy programme

S.Hugosson
National Swedish Board for Energy Source Development (NE)

Author's reply to question by Dr. M.SERT

Concerning Storage

In a fully integrated electrical supply system with good **power** reserves, the question of storing the wind energy is not very important until the total amount of installed wind power increases above $\approx 10\%$ of the total system power. This has been shown by various system studies and simulations in Sweden, USA and Germany.

The Swedish power system is supplied by hydro power to about 75% of its annual energy production. Hydro power can be regulated quite as fast as the variations in wind power would occur from large, integrated wind power stations. This requires of course, that the total supply system is integrated from a control point-of-view, which is the case in Sweden. The extra cost – to be attributed to wind energy cost – of controlling the **available** power reserves is presently estimated to be between 0:00 and 0:05 SwKr/kWh.

This cost level should apply for a total wind power installation of 3.000 - 5.000 MW in Sweden, which means approximately 1.000 large-scale WECS. The "controlling cost" should be compared to the estimated cost of wind energy **excluding** the above of 0:15-0:25 SwKr/kWh, which means that the extra cost of power balancing is reasonable.

Concerning Wind Alignment

The normal method of aligning a large WECS with the wind would be to sense the wind direction, filter its fluctuations and put the resulting signal into a yaw servo system with electric or hydraulic drive motors. This goes for both upwind and downwind turbines. Downwind turbines can be made self-aligning with servo positioning for weak winds.

Author's reply to question by Dr. D.LINDLEY

No real storage R & D concerned with the wind power problem or with utility peak levelling is performed in Sweden at present. There is no pressing need for such R & D in Sweden. I think that wind energy storage R & D should center around pumped hydro storage, as being the most compatible with **any** utility system, and also technically quite simple, being of some importance in developing countries.

For the "rural needs", no energy storage as such would be required for water pumping or refrigeration. For a small village electrical network with a few light bulbs, some radio or TV sets plus eventually a radio communication set, storage of electric energy in stationary lead-acid batteries at 98-96 V would probably be quite sufficient, technically manageable and cost competitive.

M.LODGE, INSTITUTE OF MAN AND RESOURCES, CANADA

In light of the fact that Northern Europe is not self-sufficient in food supplies, have you included in your wind energy research and development the prospect of using WECS to generate heat for direct use in greenhouses? If so, what prospects do you have or plan in this area?

Author's reply

No real effort is put into this in Sweden. There is a combined wind power - solar heating-rock storage project for a greenhouse running at the Alnarp Agricultural University outside Malmo. NE paid 50% of the wind power unit and we are awaiting test results.

Most probably, direct solar heating would be more convenient and less costly for this purpose even in Sweden.

Author's reply to question by Professor N.H. LIPMAN

In some strategic situations, the value of having an available - if fluctuating - energy, supply could certainly have a value on top of the normal commercial - energy value. This can be taken into account - through a political decision on the national level - e.g. in the form of investment or tax incentives for such forms of energy supply. The engineer developing wind power should, however, set his aim at normal commercial evaluations, **not** relying upon hypothetical - or utopian - political whims.

PAPER A3

Recent developments in wind energy

L.V. Divone
Department of Energy, U.S.A.

B. SØRENSEN, NIELS BOHR INSTITUTE, DENMARK

Regarding your cost figures:
(a) What was assumed for the cost of maintenance/operation of the wind energy generators?
(b) You compare the cost of wind-produced electricity (in fixed prices) with a fixed cost of fuel-produced electricity (e.g. 4 ¢/kWh). Are you not expecting an increase in fuel prices?

Author's reply

(a) The assumed O & M was 3% of the initial capital cost during the first year of operation. This value was assumed to increase in each subsequent year with the rate of inflation.
(b) The cost of the wind turbine system and conventional fuel prices were both escalated at a rate of 5% annually from the present time up to the analysis year (1985).

C. DRAIJER, CONSULTANT, NETHERLANDS

Taking weight as a price criterion (instead of swept area, which is generally done), my question is; is maximum torque, for a certain size, also a good criterion?
Going through a price calculation, nearly all the main parts are in size (weight) more or less proportional with maximum torque. Also the tower, the base etc. because the other principal factor, axial force is, for the same diameter/height ratio, proportional to the maximum torque.
It is not my intension to improve the price criterion mentioned by Dr. Divone, but by experience I found that maximum torque is a more reliable criterion than swept area and I see a direct connection between maximum torque and weight, all within a certain size range and for a certain type.

Author's reply

Maximum shaft torque (as an indicator of system cost) has not been used because of its great sensitivity to overall machine design. However, shaft torque is used as the dominant cost criterion for the transmission subsystem. It is an interesting idea and I appreciate the suggestion; we'll look into it further.

R.J. TEMPLIN, NATIONAL RESEARCH COUNCIL, CANADA

I have a comment about storage in large systems, and two questions for Mr. Divone about costs. The comment about storage is that although the results obtained from systems analysis are to some extent dependent on the characteristics of the particular system, it is interesting that the conclusions regarding storage are as similar for several countries (U.S., Sweden, Netherlands). In Canada also concluded some time ago from an analysis similar to that described by Mr. Divone (analysis of the effect of wind energy on the typical system load-duration curves) that storage was of little economic use to the wind part of the system, and did not appear to be economically worth while, unless it can be economically justified by the remainders of the system without wind.

The questions about costs are: (1) In Mr. Divone's chart showing costs in terms of dollars per lb weight, were the costs total installed costs, or factory production costs only? (2) Has he any **rough** feeling for the rates between factory production costs, and all of the rest of the costs involved with the installation of large systems?

Author's reply

The costs presented were total installed costs. The non-hardware associated costs of the MOD-2 early production units are expected to represent approximately 40% of the total installed cost. Although cost reductions are expected in both the hardware and non-hardware areas with increasing levels of production, it is not clear at the present time as to the relative changes that might occur.

G.R. KETLEY, BRITISH AEROSPACE DYNAMICS GROUP, U.K.

In his paper, Mr. Divone referred to the large weight of the MOD 1 machine, the blades of which are of steel. In relating the costs of this turbine to mature product cost, would he regard this structure as characterised by steel transmission towers, or large tractors, or what?

Author's reply

Asymptotic cost analyses of wind turbines, which have been performed by relating the entire wind system cost to those of "mature" mechanically similar systems, have failed to develop valid cost estimates. The present approach being used involves the estimation of mature system costs by aggregating wind turbine component costs which have been developed by comparing them with analogous products at the component level; e.g., wind turbine towers with transmission towers, generators with other generators.

J. ARMSTRONG, TAYLOR WOODROW CONSTRUCTION LTD, U.K.

The impression from the last Washington Conference is that the break through of cost effectiveness of the MOD 2 is due to allowing the resonant response frequency of tower, rotor etc to fall below excitation frequencies. Is this still the case, what are the problems, and is this principle applicable to smaller systems?

Author's reply

The MOD 2 resonant response frequency of 1.3 per rotor revolution is below the principal excitation frequency of 2 per rotor revolution. However, this resonant frequency is above the 1 per rotor revolution excitation frequency produced by rotor imbalance. A cost reduction is noted in the soft tower design which is attributable directly to this situation, although other advances also contribute to the improved economic picture.

The flexible MOD 2 soft tower design is more sensitive to extreme wind and seismic loads than a rigid truss tower design. Most of the MOD 2 tower wall thickness dimensions were designed for these extreme conditions. Therefore, no significant problems are foreseen at this point, but we still have more work to do.

This principle would seemingly offer some potential with respect to smaller systems, but to date no investigation of this application has been performed.

P. GAVA, EUROCEAN, MONACO

As the problem of onshore versus offshore wind energy systems is the cost of the energy, I think that the offshore based systems can be more cost-effective or, at least, that the problem has to be analysed further.

Author's reply

At the present time it does not appear that offshore systems will be as cost-effective as land based systems. These results are based on a recent Westinghouse Electric Co. study conducted for DOE. It does not appear worthwhile to conduct major projects on this topic in the near future, but we may continue to study it as the general technology develops.

G.R.KETLEY, BRITISH AEROSPACE DYNAMICS GROUP, U.K.

What is the minimum load into which the 200 KW MOD OA machines have been operated stably, and what is the minimum ratio between grid load and machine output which you would estimate to permit stable operation?

Author's reply

The output of the MOD-OA system in Clayton, New Mexico has never dropped below the network power load. No minimum has yet been determined.

We are presently limiting this ratio to approximately 10:1, but this may be too conservative. We will be investigating this aspect in more detail.

L.ROWLEY, CANADAIR LTD, CANADA

1. Has fatigue testing been conducted on the Kaman 150ft blade made in composite materials?

2. Are there any plans for making MOD 2 blades in composite materials?

Author's reply

1. The blade itself has not been fatigue tested. Testing has been conducted on sections cut from the manufactured prototype and 1/4 scale segment of the root bolted joint.

2. The MOD 2 contractor (Boeing) was requested to submit an alternative (TFT Composite) blade design at the preliminary design review. It was decided that this alternative type of blade did offer good potential for future applications but was not programmatically ready for implementation at this time.

D.LINDLEY, TAYLOR WOODROW CONSTRUCTION GROUP LTD, U.K.

1. The ERDA program has funded several studies of the less conventional wind energy systems such as the free vortex (Tornada) turbine the vortex augmentor etc. In Mr. Divone's presentation he refers only to the "conventional" horizontal and vertical axis types of machine Would he please comment in the light of these ERDA funded studies, on whether he sees only one of these novel wind energy conversion systems emerging as a competitor.

2. Mr. Divone has made reference to the apparent lack of difficulties in interfering WECS with utility networks, provided the total installed capacity of WECS is 'modest' and indicated that the storage issue is probably not crucial in such cases. Would he elaborate further on the evidence he has on the maximum installed capacity of WECS that can be interfaced and the dependence of this as a function of existing utility plant mix.

Author's reply

1. The Solar Energy Research Institute has just initiated a detailed investigation of this topic and expects to have preliminary results available by the Fourth Biennial Conference and Workshop on Wind Energy Conversion Systems in Washington, D.C. during October 1979.

2. The results of the New England Regional Wind Study indicated that each WECS installed has less economic value to a utility than the previous unit. This results directly from the fact that WECS replaces conventional units based upon their incremental cost of operation. From an economic standpoint, it appears that the maximum WECS penetration into a utility grid would be of the order of 30%. This value would depend on the existing utility plant mix and on whether a reoptimized mix mode or fuel saver mode was used.

S.HUGOSSON, NATIONAL SWEDISH BOARD FOR ENERGY SOURCE DEVELOPMENT

From the Swedish point-of-view we can see very small possibilities, that "tornados", "vortex augmentors" and the like could ever be competitive. The amount of material going into these designs – to achieve very moderate gains in efficiency over the stream-tube – seems prohibitive. So far, an increase in turbine diameter with a few percent would more or less do the same trick with a conventional 2-3-bladed horizontal axis turbine.

Author's reply to question by Dr. D.LINDLEY (on Storage), U.K.

Storage R&D projects are being managed by a separate entity in the Department of Energy organization. We feel that wind turbine systems do not require storage to become a viable energy supply option, but storage is vital for other reasons.

PAPER A4

The development of wind power plants in Japan

I.Ushiyama
Ashikaga Institute of Technology, Japan

P.J.MUSGROVE, READING UNIVERSITY, U.K.

In Fig.22 the author shows an asymmetric aerofoil section, and in the text (page A4-33) the author indicates that this is needed to make this giromill self-start. Can he comment on why his giromill does not self-start as a result of its cyclically varying pitch? (Cyclic pitch variation is implicit in the description of this windmill as a "Giromill")

Author's reply

In Figs.21 and 22, the author used incorrect term.
Strictly speaking this windmill is not giromill, because this windmill does not vary the pitch cyclically, namely a fixed pitch one. Now, this windmill has improved and details are shown as follows.

Specification of five square meter wind turbine generator

Wind turbine:
Diameter		2.5 m
Blade span		2.0 m
Airfoil of blades		TWT 11215-1-4012
Airfoil of arms		NACA0012
Controls		Aerodynamic Control (Spoilers on arms)
Starter		None
Gear ratio of transmission		1 : 1.4
Generator	Zephyr VLS-PM 311B	
	Rated output	1.5 Kw at 450 RPM
	Rated voltage	24 V or 48 V

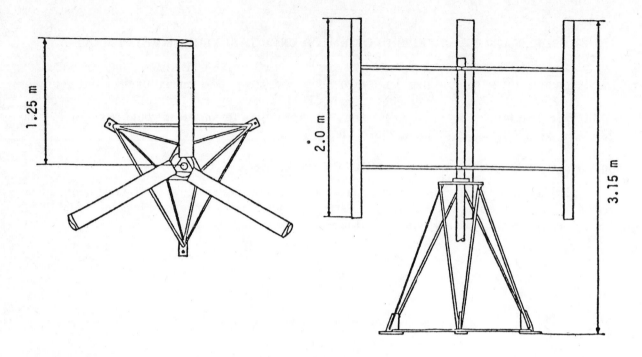

Fig. 1   General view

Performance:

(Following are at 10 m/sec wind velocity)
Turbine shaft output   1.5 Kw
Rotational speed of turbine   260 RPM
Generator output   1.0 Kw
Rotational speed of generator   364 RPM

Design wind velocities:

Cut-in velocity   3.8 m/sec
Rated velocity   10 m/sec
Maximum output velocity   12 m/sec
Cut-out velocity   15 m/sec
Gust velocity   22.5 m/sec
Ultimate velocity   60 m/sec

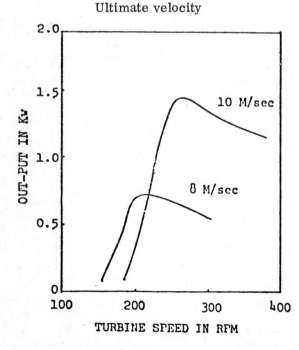

Fig. 2   Wind Turbine performance

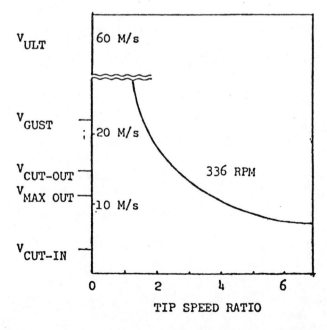

Fig. 3   Design - Wind velocity

J. M. de LAGARDE, MONTPELLIER UNIVERSITY, FRANCE

Could I ask Mr Ushiyama about the giromill of Mr Iwanaka as represented in figure 19 of his paper.

1. Are the blades of cyclic pitch type or fixed?
2. What is the type of construction for these blades?
3. Is the average position of blade perpendicular to radius of the cylinder or does it make a certain angle as it seems to appear in the figure 2.

Author's reply

First, the author show the mechanism of Iwanaka's windmill in Fig. 4. below.
1. The blades are of cyclically variable pitch type.
2. These blades are made of aluminium sheet.
3. You can see the idea from Fig. 5. below.

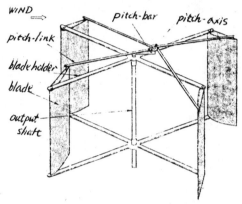

Fig. 4   Mechanism of Iwanata's mill

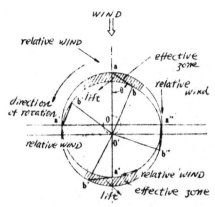

Fig. 5   Mechanism for rotation

D. LINDLEY, TAYLOR WOODROW CONSTRUCTION LTD.

On page A4-34 of his paper, Mr. Ushiyama refers to an IEA agreement concerning "the implementation of large scale wind power generation". He goes on to outline four aspects of this study; one of these is said to be concerned with the design of a 300 MW output plant, another with "the design of experiment for 3 MW component". Are these statements incorrectly printed or is it really true that the IEA have a joint project to look at a 300 MW wind energy farm comprised of 3 MW wind turbine units. Could the author (or some one familiar with the IEA programmes) elaborate.

J. PEISER, NETHERLANDS ENERGY RESEARCH FOUNDATION ECN

I wish to report that in the framework, set up by the International Energy Agency two agreements on wind energy research have been concluded:

1. Implementing agreement for a programme of research and development on wind energy conversion systems.

This agreement has four annexes:

Annex I   : Environmental and meteorological aspects of wind energy conversion systems.

Annex II   : Evaluation of models for wind energy siting.

Annex III : Integration of wind power into national electricity supply systems.

Annex IV : Investigation of rotor stressing and smoothness of operation of large-scale wind energy conversion systems.

Austria, Canada, Denmark, Germany, Ireland, Japan, the Netherlands, New Zealand, Sweden, the United States of America are participating. Two other countries are considering participation.

2. Implementing agreement for co-operation in the development of large scale wind energy conversion systems, between Denmark, Germany, Sweden, the United States of America.

M.SERT, TUBITAK-MARMARA RESEARCH INSTITUTE, TURKEY

Both speakers for papers A4 and A5 presented a survey of conventional windmills for small scale use, however a comparison of their conversion efficiencies were not made. Although wind energy appears to be free, only a limited number of machines can be used in a given area, hence conversion efficiency is important for large scale wind energy utilization, and practical results on this would be very useful.

Author's reply

Thank you for your comment. As was pointed out, the author omitted the conversion efficiency.

If necessary you can calculate the overall conversion efficiency using the equation $\eta = W / \frac{1}{2} \rho AV^3$, where A: Swept area $(\frac{1}{4} \pi D^2)$, V: rated wind speed, and W: generator output. Numerical values can be got from Table 1.

However, this is not power coefficient C_p, because W is not the windmill shaft output.

PAPER A5

Windpower programme in Tanzania

R.Reichel
University of Dar Es Salaam, Tanzania

M.SERT, TUBITAK-MARMARA RESEARCH INSTITUTE,, TURKEY

See last question to paper A4.

Author's reply

The machines in question are all small ones, thus the question of large scale wind energy utilization is hardly relevant. In developing countries and in particular in Tanzania, there is lots of space, and the end-user is not so interested in efficiency but more in price and power output, i.e. in what the machine costs and what it actually does for him. Thus the main question is: how much water is pumped daily (since water is still the priority, electricity comes later), and how long will the windmill work with no maintenance. The maintenance question is actually the most important one in remote areas!

A.J.ALEXANDER, LOUGHBOROUGH UNIVERSITY OF TECHNOLOGY, U.K.

You mention that the area of the Savonius is 10 m^2. Is this the maximum frontal area? Also the power output is given but not the windspeed or the efficiency. Do you have a figure for either of these?

Author's reply

Yes, the 10 m^2 is the total frontal area.

The rotor was already destroyed about 2 years ago, and if I remember the figures right, the one hp. output was measured at around 8 to 9 m/s wind velocity. This would mean an efficiency of about 0.2. But I am actually not very sure about these figures since I did not measure it myself.

Discussion & Contributions

SESSION B : WIND REGIMES, SITING AND INTERACTION

Chairman: J.Pelser, Netherlands Energy Research Foundation ECN.

Papers:

B1 Physical planning aspects of large-scale wind energy exploitation in the Netherlands.
A.A.Van Essen, Rijks Planologische Dienst; R.ter Brugge and J.M.van den Berg, N.V.KEMA; G.G.Piepers, Netherlands Energy Research Foundation ECN; and A.L.M.Bongaarts, Ministry of Economic Affairs, Netherlands.

B2 Program overview for the wind characteristics program element of the United States Federal Wind Energy Program.
L.L.Wendell and C.E.Elderkin, Pacific Northwest Laboratory, U.S.A.

B3 Wind energy prospecting in Prince Edward Island: A program overview and status report.
M.A.Lodge, Institute of Man and Resources, Canada

B4 Offshore wind power model estimates.
M.Garstang and R.Pielke, C.Aspliden and J.W.Snow, University of Virginia, U.S.A.

B8 Off-shore based wind turbine systems (OS-WTS) for Sweden - a systems concept study.
R.Hardell, SIKOB, Sweden, and **O.Ljungstrom,** Aeronautical Research Institute of Sweden

Chairman: O.Ljungstrom, Aeronautical Research Institute of Sweden

B5 The interaction of windmill wakes
P.J.H.Builtjes, Organisation for Industrial Research TNO, Netherlands.

B6 Wake interaction in an array of windmills. Theory and preliminary results.
T.Faxen, University of Uppsala, Sweden

B7 Measurement of wind speed around a wind power plant in Sweden
A.S.Smedman-Hogstrom, University of Uppsala, Sweden.

Contribution:

BX1 Wind characteristics research in New Zealand
D.Lindley, Taylor Woodrow Construction Ltd, U.K.

Note:

The papers were presented by the authors whose names appear in bold print.

DISCUSSION

PAPER B1

Physical-planning aspects of large-scale wind energy exploitation in the Netherlands

A.A.Van Essen
Rijks Planologische Dienst, Netherlands

R.ter Brugge and J.M.van den Berg
N.V.KEMA, Netherlands

G.G.Piepers
Netherlands Energy Research Foundation ECN

A.L.M.Bongaarts
Ministry of Economic Affairs, Netherlands

B.SØRENSEN, NIELS BOHR INSTITUTE, DENMARK

Mr. Van Essen explained, how e.g. motorways, windmill arrays and military shooting grounds under Dutch conditions would be conflicting components in planning.

Has any effort been made, to assess public preferences, e.g. between military shooting grounds and wind energy parks?

Authors' reply

This stage of the survey of the physical planning aspects of large scale wind energy exploitation in the Netherlands doesn't include public participation.

PAPER B2

Program overview for the wind characteristics program element of the United States Federal Wind Energy Program

L.L.Wendell and C.E.Elderkin
Pacific Northwest Laboratory, U.S.A.

H.SOCKEL, WIEN TECHNICAL UNIVERSITY, AUSTRIA

You made wind speed measurements on a circle. Did you calculate correlation functions between different points, because these functions are of importance for vibration problems of rotors?

Authors' reply

The analyses of these measurements to date has been of a preliminary nature. However, calculations have been made of cross-correlations of wind speed and direction measured at points opposite each other on the disc.

The wind speed cross-correlations were found to be 0.830 for the vertically-aligned pair and 0.903 for the horizontally-aligned pair. The wind direction cross-correlations were 0.539 for the vertically-aligned pair and 0.721 for the horizontally-aligned pair.

Cross-correlations as a function of time lag will be calculated to account for the effects of a rotating blade.

J.WIERINGA, ROYAL NETHERLANDS METEOROLOGICAL INSTITUTE

Mr. Wendell briefly referred to a proposal for 24-hour forecasting of winds. Could he please clarify this remark from his lecture by telling what is planned in this respect? In particular, I would like to know the degree of forecast accuracy which to Mr. Wendell's knowledge would be asked for by the wind turbine construction communities.

Authors' reply

The initial objective in this effort is to determine the reliability of operational wind forecasting methods which might be used to produce 24-hour forecasts of hourly wind speeds. Some electrical utility personnel have indicated that they would like the accuracy of wind forecasts for WECS operations to be at least as accurate as the forecasts of loads which have to be met. They have indicated that the load forecasts are reliable to within 10%.

I doubt whether current methods of wind forecasting are reliable to within 10%, but I also have some reservations about the load forecasts having this reliability. Our objective is to produce some quantitative information which will alow an assessment of the needs for, and feasibility of, development work in producing more reliable wind forecasts for this special application

J.M. de LAGARDE, MONTPELLIER UNIVERSITY, FRANCE

The watt per square meter figures quoted in the paper - were they total energy figures or energy that can be extracted taking into account the efficiency?

Authors' reply

The watt/m^2 values quoted in paper B2 represent the total power available from the wind passing through a unit area normal to the wind direction.

D. LINDLEY, TAYLOR WOODROW CONSTRUCTION LTD, U.K.

(See also contribution at the end of this session)

Dr. Wendell referred briefly to the work of Pacific Northwest Laboratory in providing data for wind turbine designers. In particular I am thinking of their very important work on estimating gust magnitudes for various values of mean wind speed and terrain roughness. From the work completed so far, is it possible to say whether,

1. Terrain has a marked effect on the detailed turbulence and gust structure of the atmospheric sheer layer.

2. Enough data presently exists to enable a rotor - tower system to be designed so that it is structurally insensitive to the effects of terrain induced variations in aerodynamic input therefore enabling a wind turbine design that is non-site dependent with respect to wind structure.

Authors' reply

Experimental evidence has confirmed that turbulence intensity increases with increased terrain roughness and complexity. However, we have contract efforts in progress which are designed to provide quantitative estimates of the effects of terrain complexity on the turbulence structure of the wind.

With evidence and experience presently available from a variety of sites, it appears that there is a very broad range of turbulence intensities and other wind characteristics pertinent to wind turbine design. The question of whether a single site-independent design would be appropriate, depends on the upper limit set on turbulence intensity and maximum winds for acceptable sites. Current projects involving the cooperative efforts of wind turbine designers and atmospheric scientists should produce results which will provide a more definitive answer to this question.

Wind energy prospecting in Prince Edward Island:
A program overview and status report

M.A. Lodge
Institute of Man and Resources, Canada

R. MATZEN, ROYAL VETINARY AND AGRICULTURAL UNIVERSITY, DENMARK

Have you details on cost and reliability for the Darrieus hydraulic pump heating system?

Author's reply

The manufacture of the turbine, Dominion Aluminum Fabricators of Mississauga, Ontario, rate the machine at 30,000 KWH annual output at a site with a mean annual wind speed of 6.2 m/s. The estimated cost of the system, including foundation, is about $10,000.

The estimate does not include starting mechanism, insulated duct, thermal storage or shipping and erection costs.

Overall system reliability is unknown. It is anticipated however that with very moderate maintanence the system should perform for at least 20 years. All parts are galvanized, stainless or otherwise protected. Bearings, pump and other hydraulic components are of high industrial quality.

G.N. NURICK, UNIVERSITY OF CAPE TOWN, SOUTH AFRICA

Would you please explain the expression on page B3-22 six lines down - that is "(4800°C days)"

Author's reply

The terminology degree-day is one commonly used in North America to describe the regional heating load of buildings. The number is determined as the sum of the product of the average daily difference between ambient temperature and 18.3 C$^{\circ}$ and the number of days for which space heating is required in the region.

PAPER B5

The interaction of windmill wakes

P.J.H. Builtjes
Organization of Industrial Research T.N.O., Netherlands

R.J. TEMPLIN, NATIONAL RESEARCH COUNCIL, CANADA

This is a comment, rather than a question. I would like to clarify two points in the introduction to Dr. Builtjes paper. The theories of Templin, Newman, etc do not make the assumption that the windmills are in lines, parallel with the wind, but only that the individual wakes of all the upstream windmills (upstream of the reference windmill) are "smeared out" into the general velocity reduction in the equilibria shear layers at the location of the reference windmill. The second point is that I do not remember having calculated an average power output as low as 0.25 W/m^2, as stated. In fact, for a spacing of about 10 diameters, for which $\lambda \approx 0.01$, Dr. Builtjes Fig. 1 shows that the theoretical infinite - field value of P/Po is about 0.5, and therefore presumably these theories would then give about 3 W/m^2 (about half of his value for a finite array of about 10 diameters spacing). The infinite-array theories are, of course, conservative, but perhaps not quite as bad a suggested. It is good to see wind tunnel measurements of the type described, however, because these will at last provide engineering data useful for cluster design, and I congratulate him on his paper.

Author's reply

I would like to comment on two points made by Dr. Templin. I agree completely with the statement that the theoretical considerations do not assume that the windmills are on a straight line, but only assume that individual wakes are smeared out. However, my experimental results show that the theoretical results compare rather well with experimental results of windmills on a straight line. In my opinion this is caused by the fact that the theoretical models are one-dimensional and consequently the theoretical results have to be compared with experimental results of windmills on a straight line.

The second point concerning the low average power output is related to the remark made by Dr. Templin in his paper prepared for the wind energy workshop at Stockholm in August 1974 that an optimum distance between windmills would be of the order of 30 diameters. This gives $\lambda \approx 10^{-3}$, and from Fig. 1, $P/Po \approx 0.9$ which leads to a value of the energy-output of the order of $0.4 \, W/m^2$.

M.SERT, TUBITAK–MARMARA RESEARCH INSTITUTE, TURKEY

Just a remark of caution on the method employed by Mr. Builtjes: To assume that the windmill wake can be simulated by the wake of a stationary object may be an experimental necessity; however, rotational nature of the windmill wake would be very difficult to reproduce; and, although the gross properties of the simulated wake may be similar, interaction of several wakes could produce different affects.

Author's reply

The wake created by the stationary object simulates the mean velocity profiles behind the turning model, and the magnitude of the total turbulent intensity, rather well. Of course, the rotational nature can not be simulated by a stationary object. However, the experiments show that the peak in the spectrum of the turbulence created by the turning of the model has disappeared after about 5 diameters.

D.J.MILBORROW, CENTRAL ELECTRICITY GENERATING BOARD, U.K.

a). What was the effective performance coefficient of your model rotors?

b). Have you investigated ground effects on your models, i.e. the reduction of output due to the proximity of the rotor to the ground?

We ask since we believe, as a result of initial results from our own tests, that these effects are significant in determining the output from a cluster of windmills. The tests are now in progress in the C.E.R.L. wind tunnel (working section 4.5m x 1.5m) using vane anemometers with a gauze fitted to the upstream face, to simulate the aerogenerators. An array installed in the wind tunnel is shown below.

The scaling of the boundary layer in the wind tunnel is at 1/500 scale and the anemometers, which are 72 mm diameter, therefore represent rotors of 36 m diameters. The gauze/rotor combination models the momentum deficit behind a windmill rotor and extracts 1.04 watts from the airstream when the windspeed is 10 m/s. This is equivalent to the energy extracted by a windmill having a performance coefficient of 0.37 and an efficiency of 87%.

Although the anemometer rotor is primarily a device for sensing the mean air velocity over the plane of the disc it does impart small quantities of swirl and turbulence to the flow. The magnitude of the swirl will probably be less than in full scale whereas the turbulence may be comparable in magnitude if not in frequency.

Since the rotational speed of the anemometer is proportional, in uniform flow, to the windspeed and since the power extracted by the gauze is proportional to the cube of the windspeed, the assembly models the performance of a "free running" windmill, i.e. one which runs at constant tip speed/windspeed ratio and efficiency. In a sheared, turbulent flow such a windmill will respond to a mean windspeed and it is assumed that the energy extraction and rotational speed of the model assemblies will continue to scale the response of a prototype machine under such conditions.

The power output which can be derived from a cluster of aerogenerators will be estimated by summing the cubes of the rotational speeds of the anemometers. The result will be compared with a reference value of total output, obtainable in the absence of interference effects. 40 anemometer/gauze assemblies are available for the tests, which will cover a number of spacing patterns, the overall objective being to maximise the power output per unit surface area.

When an array is installed in the working section the surface drag will tend to increase the height of the boundary layer; this increase can be accommodated to a certain extent but the constraint imposed by the tunnel roof may impose an artificial restraint to upward movement of the streamlines.

Since a number of assumptions have been made concerning the validity of the modelling technique, further work is in progress to verify these. Measurements on full size and model rotors and analytical techniques will aid the interpretation of the test results and enable the accuracy to be defined.

Author's reply

Question A: The grids used had a drag coefficient of 0.9.

Question B: The ground effects are investigated in so far that the grids are placed in a simulated turbulent boundary layer at 1.5 diameters above the ground and consequently ground effects are in principal included in the results. However, up till now, we have not changed the distance of the grids to the ground. A second remark is that the computer model of Dr. Hissaman (for reference see my paper) takes into account the ground effect by an image-effect. In contradiction to the suggestion made by Dr. Milborrow, in this computer model the ground effect has only a small influence consequently, this seems a very important point to investigate.

B.R.CLAYTON, UNIVERSITY COLLEGE LONDON, U.K.

1). It appears from Fig. 2 that the rotating Darrieus rotor in the wind tunnel is mounted well away from the walls of the tunnel. However, the wakes from these tests are then modelled for use with grids in a model array of turbines which are clearly in the wind tunnel shear layer. Could the author please justify the accuracy of this modelling procedure.

2). Another point raised by the author concerns the small difference between V_o and V_c some distance downstream in the wake of the rotating Darrieus rotor. This is presumably the result of turbulent mixing between the wake flow and the main stream flow over the whole cross-section of the wake about which the main stream is uniform. But is this likely if the rotor is mounted in a shear flow does to the ground?

3). Did the author make any estimates of Reynolds number effects on either the isolated wake behind the Darrieus rotor or the wake interaction experiments of the model array.

4). Did the author detect any persistant directional characteristics of the wake resulting from the deflection of the flow behind the model Darrieus rotor? If so, how were these taken into account in the formulation of the model grids? Further more, did the measurements behind the rotor reveal any strong vortex filaments propagating long distances downstream. If so, the use of a mean wind speed located at the height of the mid-point of the rotor does not seem justified.

Author's reply

Question 1. The stationary object is chosen in such a way that the object placed in a homogeneous flow, far from the walls, simulates the wake of the turning model. After that the models are placed near the ground in the way described.

Question 2. A comparison between Fig. 3, the results in the homogeneous flow, and Fig. 4, the results in the turbulent boundary layer, show a, rather unexpected, small difference between the two wakes.

Question 3. No investigation has been carried out on Reynolds number effects, save the fact that no systematic difference in model array output could be found at a velocity of 10 m/s and of 25 m/s.

Question 4. No persistant directional characteristics could be tested. For the second part see the reply to Dr. Sert.

H.SELZER, ERNO RAUMFAHRTTECHNIK, FEDERAL REPUBLIC OF GERMANY

The results of spacing the WECS are very helpful for future planning. The main interest of building WECS is to produce electricity at the lowest possible price. Therefore the costs have to be compared for the land area and for the increased size of the WECS. The ratio might become a design criteria if the land for a WECS-farm has to be paid for.

Author's reply

I agree with Dr. Selzer's remark.

PAPER B6

Wake interaction in an array of windmills-theory and preliminary results

T.Faxen
University of Uppsala, Sweden

J.WIERINGA, ROYAL NETHERLANDS METEOROLOGICAL INSTITUTE

On Mr. Faxen's paper I have serious doubts with regard to some of his hypotheses. First, the superposition of wakes by simple addition (just as if they were plumes of a passive contaminant) might lead to significant errors at close distances, say 5 to 10 radii. Second, the treatment of the ground effect by reflexion technique seems invalid in the case of wakes : closer to the ground loss of average momentum and gain of turbulence does occur in practice. Third, the used values of $\alpha = \sigma /U$ are much too low, and a range of $0.1 < \alpha < 0.2$ seems more realistic. And finally, the statement that in rough terrain wake effects are relatively less persistent is true, but its paradoxical phrasing might tend to obscure the fact that also the average wind speed is less above rough terrain. Therefore for a windmill array in rough terrain the intersection decrease will not lead to an increased power output. Unfortunately lack of discussion time does not permit me to give these comments in greater detail.

Author's reply

Dr. Wieringa has serious doubts about the model. Well, we do know that the model approach is simple but this is an absolute necessity if we want to achieve our goal: a model which is suitable as a convenient tool for the designing of wind turbine arrays, which takes all the appropiate variables into account and is physically correct. But we also believe that this is the best available at present. This belief is supported by our tests which so far show very good correlation with theory. I therefore find it a little difficult to respond to doubts when no test data is provided as support but I will do my best.

Answer to question:

1). It is assumed in the model that the windmills should be located so that no wake interaction will occur in the first 10 radii downstream of each unit.

2). The loss of average momentum due to ground friction is not accounted for in the model since this is included in the shape of the oncoming wind profile defined by the upstream meteorological parameters and ground roughness. We therefore assume a steady state wind profile and that these conditions do not change within the array.

Further on, yes the turbulence intensity is higher close to the ground but this does not change the behaviour of the wake since the ground also acts to reduce and eliminate the turbulent transport of momentum deficit and the mixing with outer flow. Thus wake velocities are smaller due to the presence of the ground which is what we get with the image model. The only condition here is the rotor must not be too close to the ground.

3). The value of α now used is in the range of 0.1.

B.R.CLAYTON, UNIVERSITY COLLEGE LONDON, U.K.

1). Since the rotor of most wind turbines has few blades the flow in the downstream wake must be both non-uniform and unsteady. Could the author explain how in his analysis the unsteady nature of the flow, caused by blade passing, is taken into account since similarity with jet flows is inapplicable.

2). Is the analysis presented by the author applicable to the vertical axis rotor bearing in mind that the wake is not symmetric in the wind direction?

3). Is it possible for the author's analysis to incorporate the effect of Reynolds number so that a scale effect can be taken into account when transferring theoretical predictions to model and prototype arrays?

Author's reply

First I would like to refer to my statement in the answer to Dr Wieringa's question.

1). The important thing is the loss of momentum and that this momentum deficit is constant throughout the wake. This together with the turbulence intensity controls the shape of the mean wind profile. A few diameters downwind, the strong turbulent mixing will result in a flow which is steady in the mean. Non uniformity is specifically not assumed and is incorporated in the wake profiles.

2). Yes, downwind both devices have a wake like an energy extracting drag device.

3). Reynolds number does not occur specifically here. The important items are the scale and intensity of turbulence. This is specifically taken into account in the turbulence intensity parameter α.

G.W.W.PONTIN, WIND ENERGY SUPPLY CO. LTD., U.K.

Could the author confirm, in view of the rotometric formulae quoted in the paper, that familiar amplitude effects are not neglected. In particular, are Reynolds number effects included or not?

Author's reply

Turbulent entrainment at those scales is normally assumed to be independant of Reynolds number, that is, it is in the Inertial Mixing Range (as shown e.g. in Monin and Yaglom 1973, Part I, p. 351).

Here I also refer to my answers to Dr. Wieringa's questions.

PAPER B7

Measurement of wind speed around a wind power plant in Sweden

A.S.Smedman-Hogstrom
University of Uppsala, Sweden

G.W.W.PONTIN, WIND ENERGY SUPPLY CO. LTD., U.K.

In view of the statistical methods necessary and being employed, would someone or somebody standardise averaging times (as used for wind speed measurements) as a matter of some urgency?

Author's reply

I don't think there is any need for a standard averaging time because the averaging time depends on the distance between the windmill and the anemometer.

C.J.CHRISTENSEN, RISØ NATIONAL LABORATORY, DENMARK

Dr Smedman-Hogstrom's remarks on coherence between wind velocities at different points in space are very important to stress. Aerodynamic models of windmill rotors usually use a laminar wind stream of constant velocity and they tend to ask the meteorologist for instantaneous values of wind velocity and power production.

To answer this question is to violate the laws of nature because of the stochastic nature of turbulent disturbances. As you always have a distance from the anemometer to the windmill, Dr Hogstrom's exponential expression for coherence shows that only for low frequency disturbances you can expect a deterministic connection between measured wind speed and power, whereas high frequency disturbances are different at the two points.

This means that if one tries to measure instantaneous values or short time averages, one is bound to get power curves with heavy scatter. One is simply not measuring the same wind that drives the mill.

Author's reply

It is quite right that one should measure wind speed as an average value over quite a long time (60 sec or more) when the distance between the wind mill and the anemometer is about 100m.

Off–shore based wind turbine systems (OS–WTS) for Sweden–
a systems concept study

R.Hardell, SIKOB, Sweden
O. Ljungstrom, Aeronautical Research Institute of Sweden

P.J.MUSGROVE, READING UNIVERSITY, U.K.

Could the authors indicate
i). The cost of the offshore tower/foundation structures for water depths of the order 20 to 30 m.
ii). The spacing between adjacent windmills in their offshore groups.
iii). The power rating of the cable, for which they quoted a cost of 1.3 million Kroner for transmission over a distance of 25 km.
Could the authors also indicate the load factor (i.e. average output relative to the rated output) for the large Swedish wind turbine designs.

Authors' reply

i). The cost of the tower and foundation structures is of the order of 3 mill. Sw. crowns.
ii). The spacing at 6-7 MW capacity is 800 m to 1000 m.
iii). The 130 kV AC cable transmits 100 MW of electric power.
The load factor for the large Swedish wind turbine designs is 0.28 to 0.32.

C.W.M.TILENIUS, CONSULTANT, FEDERAL REPUBLIC OF GERMANY

Please explain your envisaged solutions to the following problems in your designs:
1). Deicing of the blades of the rotors
    a) before or at the start-up phase;
    b) while in operation
       to keep this rotor unbalance within safe limits e.g. limit the breaking off pieces of ice to maximum small sizes.
2). Protection against lightning
3). Protection against changing electrostatic charges on the rotor blades during rotation through the earth potential resulting in radio interferences

Authors' reply

1). Icing is a problem for some 4 days per annum in some locations on land - perhaps more days at sea sites. Deicing is not required in the Swedish MW size WTS design spec., but ice detection devices are recommended. There is an operating design case with unbalance due to ice shed from one blade (max 15% of blade weight)
2). The Swedish design specs require lightning protection from rotor tip via hub-nacelle-tower to ground, e.g. avoiding stroke current passing through bearings (slip rings).
3). No definite solution, problem will be watched.

B.SØRENSEN, NIELS BOHR INSTITUTE, DENMARK

Can you tell us the expected lifetime of the fundament structure in water?

Authors' reply

The expected lifetime of the fundament is more than 30 years.

J.WIERINGA, ROYAL NETHERLANDS METEOROLOGICAL INSTITUTE

I have a question on bird migration. I am no expert on this subject, but during the preparatory stage of the Netherlands' wind energy program (as exposed in Mr. Piepers' paper earlier during this conference) we looked into a large number of possible constraints on windmill exploitation. Concerning bird migration the biological experts informed us that the preferred migration routes were along coastlines at a few Km distance from it and at low heights. Since it is known that major bird migration occurs along the Southern tip of Skåne in particular, there is a slight possibility that situation of a large amount of high windmills at that site would be ecologically murderous. Could one of the authors please comment on that problem?

Authors' reply

Bird migration problems are of course being watched carefully in the Swedish program and as regards offshore sites, I would prefer to locate WECS some 2-3 Km. from the shore, which is quite feasible in the Skåne coast areas. At night, I propose lighting the rotor shell to make the slow moving rotor visible at close range. In daylight, MW size rotors should be easy for birds to avoid.

CONTRIBUTION BX1

WIND CHARACTERISTICS RESEARCH IN
NEW ZEALAND

D. Lindley
Taylor Woodrow Construction Ltd, U.K.

I would like to bring to the attention of delegates some work supported by DOE through Pacific Northwest Laboratories and mentioned briefly by Dr. Wendell in his oral presentation of the paper. From July 1977 to January 1978, Professer Bob Meroney of Colorado State University, Civil Engineering Department, was assisted by DOE in a joint program of work with our small team in the Department of Mechanical Engineering at University of Canterbury Christchurch, New Zealand. During this period, wind tunnel measurements were combined with local climatological measurements and a mobile field program to assess site locations for Wind Energy Conversion Systems. The program aimed (a) to utilise simulation of the relevant wind characteristics in a meteorological wind tunnel (b) evaluate the validity of laboratory simulation methods and (c) provide a confidence measurement bound for laboratory data by the simultaneous completion of a limited field measurement program.

The laboratory method consisted of obtaining velocity and turbulence measurements over a scale model of the selected terrain placed in a simulated atmospheric flow. The area studied is located along the Rakaia River as it emergies from the Southern Alps in the South Island of New Zealand. A model section representing an area 6100m wide x 18,300m long was constructed to a scale of 1 : 5000 - various model parameters were studied and included the effects of 'terraced' versus 'smooth' models and shelter belt and vegetation simulation techniques. The Field experiment called for measurements of wind velocity and direction over the test region on a day having a north westerly stable, strong wind event. Three teams took ten 15 minute wind records each at 10m above ground level, over a 5 hour period on two different days in locations previously fixed during the wind tunnel excercise. The measurements were normalised against continuous records taken from long term anemometer stations in the same region. The results of those studies have been reported in full (Refs 1 and 2 below),; the major conclusions for this experiment are as follows. Physical modelling can:-

1. Reproduce wind patterns produced by the atmospheric shear layer flowing over complex terrain to within the inherent variability of the atmosphere to produce stationary results:

2. Reproduce relative wind speeds found over complex terrain by rank to sample correlation coefficient levels equal to 0.78 to 0.95

3. Reproduce the individual day to day quantitative wind speeds found over complex terrain to sample correlation coefficient levels equal to 0.7 to 0.76.

4. Reproduce the two field day average quantitative wind speeds found over complex terrain to a sample correlation coefficient level equal to 0.81.

5. Reproduce the individual day to day site wind directions found on complex terrain to sample correlation coefficient levels equal to 0.65 to 0.67.

The work further showed that :-

6. Adequate physical modelling of adiabatic shear flow over complex terrain requires attention to surface roughness, terrain shape, and vegetation as well as upstream velocity profile, turbulence intensity, and turbulence eddy structure.

7. Over complex terrain local wind speeds may vary by over 100% in a distance of a few hundred meters as a result of terrain shadowing, flow separation, or flow enhancement,

8. In the Rakaia River Gorge area the preferred WECS locations were the surrounding hills and ridges and not the gorge or river bottom.

This work is being extended at the University of Canterbury and Colorado State University to look at other classes of terrain. The work in New Zealand is complementary to the work of the New Zealand Wind Energy Task Force (funded by the New Zealand Energy Research and Development Committee) which is presently extending its network of over 60 wind energy resource measurement stations to over 100 such stations. They are investigating spatial variations in wind energy flux as well as vortical velocity profiles, short and long term wind speed variations, and maximum gust speeds. Details of the Task Force study are available from

Dr. N. Cherry, Dept of Agricultural Engineering, Lincoln College, Canterbury, New Zealand.

References

1. Meroney, R.N.
    Bowen, A.J.
    Lindley, D.
    Pearse, J.R.
    "Wind characteristics over complex terrain : Laboratory simulation and field measurements at Rakaia Gorge, New Zealand, New Zealand Energy Research Development Committee (Report), Univ of Auckland. July. 1978.

2. Meroney, R.N.
    "Prospecting for Wind Energy". International Solar Energy Society Annual Meeting Aug. 28-31, 1978 Denver, Colorado.

Discussion & Contributions

SESSION C : DESIGN AND CONSTRUCTION : GENERAL

Chairman: R.J.Templin, National Research Council, Canada

Papers:

C1   Atmospheric turbulence structure in relation to wind generator design.
N.O.Jensen and S.Frandsen, Risø National Laboratory, Denmark.

C2   Performance evaluation of wind power units.
O.A.M.Holme, SAAB-SCANIA AB, Sweden.

Chairman: H.Selzer, ERNO Raumfahrttechnik GmbH, Federal Republic of Germany.

C3   Investigations on the aeroelastic stability of large wind rotors
H.H.Ottens and R.J.Zwaan, National Aerospace Laboratory, Netherlands.

C4   Aeroelastic Stability and response of horizontal axis wind turbine blades.
S.B.R.Kottapali and **P.P.Friedman,** University of California, U.S.A. and A.Rosen, Technion – Israel Institute of Technology, Israel.

C5   The aeroelastic behaviour of large Darrieus-type wind energy converters derived from the behaviour of a 5.5m rotor
A.J.Vollan, Dornier Systems GmbH, Federal Republic of Germany

C7   The effect of control modes on rotor loads
E.A.Rothman, Hamilton Standard, U.S.A.

C6   Gears for wind power plants
P.Thornblad, Stal-Laval Turbin AB., Sweden

Notes:

The papers were presented by the authors whose names appear in bold print.

ERRATA

PAPER C3

Page C3–31   Delete the last line of the summary beginning with
"Finally difficulties....."

Pages C3–47   Delete Figures 14, 15 and 16.
    C3–48

PAPER C6

Page C6–93   In the table,
column 3, line 1 should read "530–700" (not "53–700")
column 2, line 4 should read "670–780" (not "67–780")

DISCUSSION

Atmospheric turbulence structure in relation to wind generator design

N.O.Jensen and S.Frandsen
Risø National Laboratory, Denmark

H.SOCKEL, WIEN TECHNICAL UNIVERSITY, AUSTRIA

You considered only the along wind fluctuation component of the wind. But if you make a straight forward first order theory, I think the cross wind turbulence component is of the same order as the along wind one for the problem.

Authors' reply

I must refer you to the arguments put forward in the appendix.

P.BUILTJES, ORGANISATION OF INDUSTRIAL RESEARCH TNO, NETHERLANDS

The turbulence generated in a field of windmills will perhaps determine the minimum required distance between windmills and make it larger than the 6 diameters mentioned in my paper.

Is there a possibility of incorporating the turbulence produced by a first windmill on a second windmill in your calculations?

Authors' reply

In principle that would be possible. It would require knowledge about intensity, spectral composition, and spatial structure of the turbulence in the wake of the first windmill, however. Whether considerations of this sort would determine the minimal distance between mills I can't say, but general knowledge about jets, wakes, plumes, thermals, and internal boundary layers say that: the turbulence very quickly establishes equilibrium (inertial subrange) and decays away in a predictable fashion that the rate of spread is about 20°, and that the mean flow reaches equilibrium before the turbulence level does. At a downwind distance of 6 d there would remain of the order of 10% of the turbulence introduced by the first mill.

P.FRIEDMANN, UNIVERSITY OF CALIFORNIA, U.S.A.

I have a comment regarding your final conclusion on page C1-9 "It is concluded that turbulence must be taken into account in wind power designs. In principle this is possible through the methods indicated in this paper".

First the method you use is the conventional method used in the field of evaluating wind effects on building, where linear models and constant coefficient dynamic systems are usually considered to be adequate. The rotor dynamic response problem however is governed by equation with periodic (time dependent) coefficients. Therefore the conventional power spectral density approach, advocated in your paper, is not valid.

Furthermore operating experience gained with the NASA ERDA MOD-O machine indicates that the response is deterministic occurring at multiples of the rotor speed $1, 2/\Omega$. Therefore it seems that deterministic sinusoidal gust models might be adequate.

Would you care to comment?

Authors' reply

First of all it should be emphasized that it was not our intention to present an actual rotor design model but only to show that turbulence is of equal importance in the design of wind energy generators, as it is when dealing with tall building structures.

Your point that lift and drag coefficients are time dependent is true for fast variations in the angle of attack. This would be important for turbulent eddies with wavelengths less than of the order of 10 cord widths. Refering to our Fig. 5 and eq. (18), the fraction of variance fulfilling this criterion is only $\sim 4\%$ of the total variance in this particular example. Hence, it is not necessary to consider time dependent coefficients in the turbulence buffeting problem the paper is dealing with.

Regarding a deterministic variation in the loading caused by the working of the rotor in a mean wind shear, the length of a sweep from the top to the bottom position of a blade is much more than 10 cord widths for any load giving part of the blade. Thus, the assumption of quasi-stationary conditions is again in place. Strong response at frequencies corresponding to the rotor revolution frequency is probably caused by the combined effect of the mean wind shear and the variations in gravity load, and is not related to the gust loading problem.

Your deduction that a sinusoidal input probably will suffice because the observed loads look sinusoidal is erroneous. The process might be of a very narrow band width as in the case of low damping, so that the response seems to be nearly harmonic, but it is never the less random.

PAPER C3

Investigations on the aeroelastic stability of large wind rotors

H.H.Ottens and R.J.Zwaan
National Aerospace Laboratory, Netherlands

D.J.SHARPE, KINGSTON POLYTECHNIC, U.K.

For very large VAWTS gravitational loads increase in significance. Were they accounted for in the geometrical stiffness matrix?

Authors' reply

No, we did not account for the gravitational loads in the geometric stiffness. In our formulation the geometric stiffness was taken proportional to the operational speed squared, while the contribution of the gravitational load is independent of the rotation. Of course, this effect can be accounted for using two different geometric stiffness matrices.

PAPER C4

Aeroelastic stability and response of horizontal axis wind turbine blades

S.B.R.Kottapali and P.P.Friedmann
University of California, U.S.A.

and

A.Rosen
Technion-Israel Institute of Technology, Israel

P.C.HENSING, DELFT UNIVERSITY OF TECHNOLOGY, NETHERLANDS

You assume in your paper that the angular blade speed Ω is constant. Due to the in-plane bending deformations of the blades the angular blade speed will slightly vary. Do you neglect this "rigid body motion" and do you have an impression in that case of the influence of this motion on the calculated results?

My second question concerns the neglection of the blade torsion. In most of the aeroelastic studies on aircraft torsion this turns out to be very important. For this reason the elastic axis and the quarter chord line of helicopter blades coincide very often. The torsional frequency of a rotor blade must be very high in order to allow the neglection of blade torsion. What is your comment on that?

Authors' reply

The assumption of uniform angular speed of rotation Ω and in-plane bending degree of freedom are two completely independent items. Since the in-plane elastic deformation of the blade is represented by V (lag deformation) the inertia terms associated with this degree of freedom account, in an exact manner, the dynamic effect indicated.

In reply to the second question: Since blade torsion is included in the analysis the question is irrelevant.

PAPER C5

The aeroelastic behaviour of large Darrieus type wind energy converters derived from the behaviour of a 5.5m rotor

A.Vollan
Dornier Systems GmbH, Federal Republic of Germany

D.J.SHARPE, KINGSTON POLYTECHNIC, U.K.

For very large VAWTS gravitational loads increase in significance. Were they accounted for in the geometrical stiffness matrix?

Author's reply

The gravitational loads give rise to a geometric stiffness term which is independent of the rotational velocity. The influence of this additional term was not accounted for in the present paper.

However, including the gravitational stress (with a maximum of $100 \, N/mm^2$) in the modal analysis of a relatively heavy aluminium blade for a 50m Darrieus-rotor, the influence of this geometric stiffness term on the frequencies were found to be less than 0.5%. This together with the only minor changes in the mode shapes, would not alter the critical velocity significantly.

PAPER C6

Gears for wind power plants

P.Thornblad
Stal-Laval Turbin, A.B., Sweden

D.J.SHARPE, KINGSTON POLYTECHNIG, U.K.

Flexing the anulus gears must cause high stress concentrations at the roots of the teeth. Does this give rise to a severe fatigue problem?

Author's reply

Flexing of the internal gear rings causes bending stresses in the rings, and they effect the stress pattern in the root section of the teeth. This influence is, however, rather small in comparison with the stress caused by bending of the teeth.

Flexible rings is not a feature which is new to this design. Similar flexibility is utilized in the Stoekicht design.

More than 800 such gears are used in marine propulsion machineries of the type referred to in the paper (Figure 14), and they have together accumulated more than 30 million hours of service. No case of fatigue problems due to ring flexibility has come to our knowledge.

I.D.MAYS, READING UNIVERSITY, U.K.

In your paper you refer to a 2.5MW system:
Do you have any tentative costs for a machine?
Your machine has been designed for application to a horizontal axis wind turbine. Is there any basic problem in applying it to a vertical axis machine, by turning the gear box through 90°?

Author's reply

I have preferred to give costs in relative terms rather than in absolute, because the price is dependent on many different factors, which have to be evaluated from case to case. Most important of these is the number of gears. A price quoted for a single unit is quite different from prices for series production in large numbers. The latter is the valid figure for investigations of future economics of wind power utilization, while the former is of interest for making cost estimates for a prototype plant.

The gear shown in Figure 15 of the paper is intended for a horizontal axis wind turbine. With slight modifications of design details it can equally well be used for vertical axis machines.

F.H.THEYSE, THEYSE ENERGIEBERATUNG, FEDERAL REPUBLIC OF GERMANY

I appreciate very much Mr. Thornblad's presentation of the advantages of epicyclic gears.

His examples beautifully design around the relevant Stoekicht patents ext, but give insufficient attention to the torsional deformation of specifically the sun-wheels.

Further I should like to draw attention to the power and speed control properties imported onto power transmission systems, using the ring-wheel in epicyclic gears, driving or braking them.

Author's reply

The torsional deformation of the sun wheel is only of some importance in the third stage (the high speed end), where the sun wheel is rather slender. Through grinding of the teeth of this sun wheel with suitable modifications we compensate for the torsion.

We have briefly looked into some alternative methods to use rotation of the gear stator so that the wind turbine rotor speed can be allowed to vary slightly although the alternator speed is kept constant. These studies have not been deep enough to establish if it is advantageous from a cost point of view to adopt such a system and instead make the blade governing system simpler and cheaper. We appreciate that these possibilities exist with epicyclic gear transmissions, and it may very well be that this feature will turn out to be another important factor in favour of selecting epicyclic gears for wind power plants.

Discussion & Contributions

SESSION D : DESIGN AND CONSTRUCTION : HORIZONTAL AXIS TURBINES

Chairman: S.Hugosson, National Swedish Board for Energy Source Development

Papers:

D1 Test results from the Swedish 60 kW experimental wind power unit
B.Gustavsson and G.Tornkvist, SAAB–SCANIA, Sweden

D2 Measurements of performance and structural response of the Danish 200 kW Gedser windmill.
P.Lundsager, Risø National Laboratory, Denmark; V.Askegaard. Technical University of Denmark; and E.Bjerregaard, Danish Ship Research Institute.

D3 Design concept for a 60m diameter wind turbine generator
D.F.Warne, Electrical Research Association, U.K.; G.R.Ketley, British Aerospace Dynamics Group, U.K.; D.H.Tyndall, The Cleveland Bridge and Engineering Co. Ltd., U.K.; and R.Crowder, Taylor Woodrow Construction Ltd., U.K.

Contribution:

DX1 Measurements of performance and structural response of the 200 kW Gedser windmill. Status medio September 1978.
P.Lundsager, Risø National Laboratory, Denmark; V.Askegaard, Technical University of Denmark; and E.Bjerregaard, Danish Ship Research Institute.

Note:

The Papers were presented by the authors whose names appear in bold print.

DISCUSSION

PAPER D1

Test results from the Swedish 60kW experimental wind power unit

B.Gustavsson and G.Tornkvist
SAAB-SCANIA, Sweden

A.J.ALEXANDER, LOUGHBOROUGH UNIVERSITY OF TECHNOLOGY, U.K.

In Fig. 8 of the paper the velocities close to the ground seem high. Can you say what conditions were like up stream? For example did trees distort the velocity profile?

Authors' reply

The terrain changes from open country to rather high wood in the direction 40°. This can be seen in Figure 1, Paper B7.

In my paper typical hourly mean profiles are shown for wind directions on both sides of the terrain change.

W.A.M.JANSEN, STEERING COMMITTEE ON WIND ENERGY FOR DEVELOPING COUNTRIES, NETHERLANDS

Why is predicted output of your 60 kW machine 20% higher than measured output? See Fig. 5 Paper B7 how do you explain this?

Authors' reply

The theoretical curve for the turbine power in Fig. 5 of paper B7 was computed with a Reynolds number 6 000 000 according to Abbot, J.H. and van Doenhoff, A.E. Theory of wing sections, Dover Publications Inc, New York 1959 and adapted to a turbulent boundary layer. The high Reynolds number was thought to be on the safe side for stress calculations. Later power has been calculated with a lower more appropriate Reynolds number 1 500 000 and losses due to coning and machinery has been considered in contrast to the earlier case.

The calculated power together with the measured values are shown in a new Figure (See below). As can be seen in the Figure the calculated curve lies entirely in the area of the measured values. Compared with the regression analysis curve there is a slight difference. A plausible explanation of the difference at low wind speeds is higher losses in the machinery than 5 percent which was used. The low calculated values at high wind speeds can owe to a calculated exagerated influence of Reynolds number or changed stalling characteristics compared with two dimensional flow.

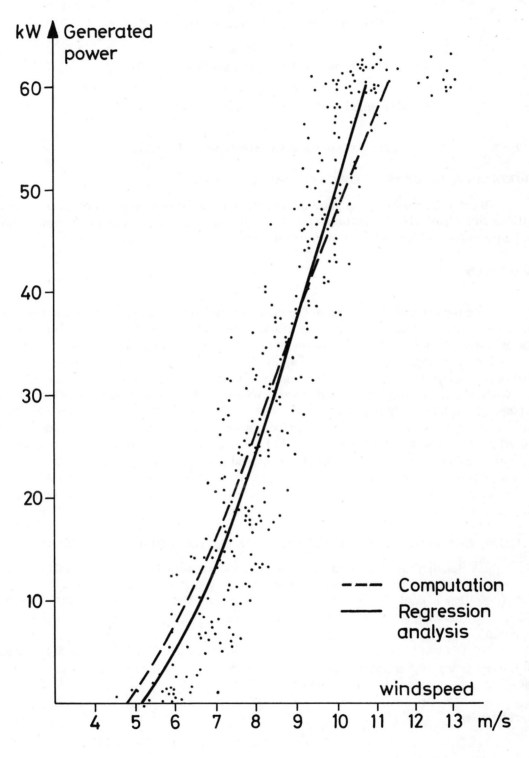

Fig. 7

PAPER D3

Design concept for a 60m diameter wind turbine generator

D.F.Warne
Electrical Research Association, U.K.

G.R.Ketley
British Aerospace Dynamics Group, U.K.

D.H.Tyndell
The Cleveland Bridge and Engineering Co. Ltd., U.K.

R.Crowder
Taylor Woodrow Construction Ltd., U.K.

P.FRIEDMANN, UNIVERSITY OF CALIFORNIA, U.S.A.

Have you considered in your design the structural dynamic penalty associated with variations in torque? If so, what analysis have you used to evaluate coupled rotor/tower aeroelastic response due to high torque variations?

Authors' reply

The design calculations thus far have not included an evaluation of the coupled rotor/ tower aeroelastic response to torque variations, although it is acknowledged that they should do so before the construction phase is commenced. There is some reason to believe that, with this type of machine in which the variation is applied torque with gusting is inherently limited by the stalled operation of the turbine over the higher wind speed range, the relative amplitude of the torque fluctuations may be less than that experienced with, say, the NASA machines. Nevertheless, it is clearly important to verify that no natural mode of the coupled rotor/tower system is resonant with a multiple of rotational speed, and that all modes are adequately damped. Calculations which have been made of isolated turbine natural frequencies indicate that these conditions are likely to be satisfied by the proposed design. It may also be noted that the use of an induction-type generator running at several percent slip, instead of a synchronous machine, will provide greater damping due to the speed variation which occurs in response to gusting, again reducing susceptibility to large transient torques.

H.SELZER, ERNO RAUMFAHRTTECHNIK, FEDERAL REPUBLIC OF GERMANY

1. Running the turbine at variable speed at most of the time, you will encounter problems of resonance between rotor and tower. Did you look for this?
2. The annual energy output diagram showed values of about 12 GWh up to windspeed of 15 m/s. This means at least 4000 hours per year of 15 m/s. To my knowledge there must be something wrong.
3. Did you make cost considerations? Running the WEC up to 3.9 MW in order to catch the few hours of very high wind speed what are the extra costs for the increased loads, weights etc in comparison to a rated power design?

Authors' reply

1. The questioner misunderstands the mode of operation of the machine - it runs at near-constant 33-34 r.p.m.
2. The diagram referred to included calculations up to 15 m/s average windspeed in order to ensure that correct shape of the curve at values below this. It is acknowledged that there are no known sites at which the wind speed actually reaches this value.
3. Cost comparisons with lower rated power designs employing variable pitch or other means of power limitation have indicated that, for high mean wind speed sites, the additional electrical output generated at the high end of the wind speed regime, together with the saving in

cost due to the simpler and more reliable turbine and control mechnism, outweigh the extra cost of the larger generator and stronger structure and transmission. The increase of design loads due to the higher-rated generator is, in fact, small since the variable pitch mechanism response cannot be made sufficiently fast to alleviate the transient loads due to gusts to the same extent that it alleviates the steady-state load, and both types of machine must therefore be designed to accept substantially the same transient loading. It is acknowledged that, for low average wind speed sites, the cost comparison may move in favour of a more sophisticated power control system, but such sites are unlikely to be economic whatever form of machine is used.

R.HARDELL, SIKOB, AB, SWEDEN

In your paper you remark that you have examined several design parameters, e.g. variable pitch for their influence on annual energy output and on specific energy costs.

Is this statement concerning costs valid for the prototype only, or even for series production of wind turbines?

Authors' reply

Our study concentrated primarily on minimising the costs of small numbers of machines, since our terms of reference were concerned with the costs to be incurred in commencing wind power exploitation during, say, the next decade, during which time large scale production would not be warranted. Also, it was concluded from other studies that predictions of the cost reduction which might be achievable from large scale production were highly speculative and of low reliability, and should not be allowed to over-influence the choice of design for initial experimental units. When several years successful operation of a small number of machines has been accomplished, the way will be clear to evaluate accurately the cost reductions from system or structural design modifications to facilitate series production.

F.J.C.SCHELLENS, FDO TECHNISCH & ADVISENTS BV, NETHERLANDS

In relation to the rotor diameter of 60m the installed power of 3.9 MW is a very high figure.

For Dutch coastal wind conditions and a turbine with a rotor blade pitch angle device we should use a design windspeed of 12 to 13 m/sec, resulting in an installed power of about 1.2 MW. With a rotor, like in your design, with fixed rotorblades it would be advisable to use an aerodynamic blade profile with a sharp $c_p - \lambda$ curve.

This lowers the investment in rotating equipment and the energy output could even be higher due to better overall part load efficiencies in the more prevailing windspeed regions.

Authors' reply

Reply not received at time of publication.

MEASUREMENTS OF PERFORMANCE AND STRUCTURAL RESPONSE
OF THE DANISH 200 kW GEDSER WINDMILL

STATUS MEDIO SEPTEMBER 1978

P. Lundsager
Risø National Laboratory, Denmark

V. Askegaard
Technical University, Denmark

E. Bjerregaard
Danish Ship Research Institute, Denmark

This contribution is an update of Paper D2, published in Volume I of these Proceedings.

Table 2a contains the updated schedule for the measurement program. During the spring campaign 7 runs were made at average wind speeds up to 12 m/s and instantaneous wind speeds up to 18 m/s. During these runs a 95% availability of the sensors were obtained. During the summer, preliminary data processing was made, mainly consisting in the conversion of data from a selected run to physical units, and spot checks on these data were made. The work done so far is described in the interim report (ref. 6). The start of the autumn campaign has been postponed by approx. $1\frac{1}{2}$ months due to repair work on the transmission of the mill.

Preliminary results

The preliminary results listed here are extracted from the interim report ref. 6. The plots figs. 9 - 12 show records from a short run including start and stop of the mill. The symbols used refer to table 6 and fig. 8. This is the run used for spot checks. The average wind speed was approx. 12 m/s.

The most important results, based on the preliminary data processing, are listed below:

The quality of the records

-   The records are generally sufficiently detailed for meaningful frequency analysis. Fig. 12 shows details of some of the records, and fig. 13 shows the result of a preliminary frequency analysis of the electric power output.
-   The records are almost free of noise.
-   The spot checks indicated very few channel malfunctions. However, a small number of channels suffer from calibration problems.

The interpretation of the records

-   During spot checks some problems became apparent, primarily:
-   Unknown external forces during zero runs may cause problems in the interpretation of results from the rotor sensors (groups 1 and 3) and the measurement cylinder gauges (group 4).
-   The calculation of external forces from the measured data demands very detailed and accurate modelling of the structure. Direct measurements of the blade deflections would be of great help.

The behaviour of the mill

-   Both average stresses and stress amplitudes in the rotor assembly are small, compared to the design stress 59 MN/m^2 (600 kp/cm^2). In table 7 representative values for ~ 12 m/s are listed.

-   The spot checks indicate that all three blades extract an equal amount of power from the wind. Due to internal stay forces, however, the amount of mechanical power delivered from each blade seems to deviate more than 30% from the aerodynamic power.

-   Both average forces and force amplitudes in the rotor assembly are listed in table 8. The outer stay forces NYS carry most of the wind load, while the outer wire forces NYB-31 and NYB-32 carry most of the gravity loads on the blades. Thus the stays and wires actually play a significant role in the distribution of forces in the rotor assembly.

-   The pressure transducers indicated a stall transition zone at radius 5 to 7 meters

-   Acceleration and yaw rate levels of the nacelle are low. Gyral forces seem to be low and the run of the mill is very smooth.

-   Fig. 14 shows the power curve and efficiency of the mill as determined from 10 min. averages from the long term measurements. The windspeeds are too low to show the power cutoff due to stall. The peak efficiency appears at 8-9 m/s.

-   Fig. 13 shows that a pulsation frequency of about 0.7 Hz in the electric power output is still present. The phenomenon was detected early during the operation of the mill (ref. 4), and a study of the phenomenon is going on in order to supplement investigations described in ref. 4. The power fluctuations, having amplitudes up to ~ 40 kW, most probably are due to torsional vibrations of the rotating parts excited by wind turbulence, and the coupling of the asynchronous generator to the grid may be a significant parameter.

Reference

6. Lundsager, P., Christensen, C.J., Frandsen, S. (ed.)
    Interim report on the Measurements on the Gedser Windmill
    Report GTG 771.105-1 Sept. 1978.

Discussion & Contributions

SESSION E : DESIGN AND CONSTRUCTION : VERTICAL AXIS TURBINES

Chairman: L.Divone, Department of Energy, U.S.A.

Papers:

E1   Operating experience with the Magdalen Islands wind turbine
P.South, R.Rangi and R.J.Templin, National Research Council
of Canada.

E2   A 200 kW vertical axis wind turbine-results of some preliminary tests
J.H.Van Sant, R.D.McConnell and **A.Watts,** Hydro Quebec Institute
of Research, Canada

E3   The design, construction, testing and manufacturing of vertical axis
wind turbines
R.Braasch, Sandia Laboratories, U.S.A.

E4   Development of the variable geometry vertical axis windmill
P.J.Musgrove and I.D.Mays, Reading University, U.K.

E5   A new concept: the flexible blade windmill (FBW)
J.M. de Lagarde, Montpellier University, France

E7   The cycloturbine and its potential for broad applications
H.Meijer Drees, Pinson Energy Corporation, U.S.A.

E6   Wind tunnel corrections for Savonius rotors
A.J.Alexander, Loughborough University of Technology, U.K.

Contributions:

EX1   Vertical axis windmill development by British Aerospace
A.C Willmer, British Aerospace, U.K.

EX2   A new vertical-axis wind turbine (summary of film presentation)
L.Arnbak, Scangear Ltd, Denmark

EX3   Experimental investigation on the performance of a Darrieus-rotor
in natural wind.
F.Rasmussen and B.Maribo Pedersen, Technical University, Denmark

EX4   Practical considerations in the design of a vertical-axis windmill
F.C.Evans, University of St. Andrews, U.K.

Notes:

The papers were presented by the authors whose names appear in bold print.

Paper E3 was presented by E.G. Kadlec of Sandia Laboratories, U.S.A.

Paper EX1 was presented by P. Musgrove of Reading University, U.K.

DISCUSSION

PAPER E2

A 200 kW vertical axis wind turbine—results of some preliminary tests

J.H.Van Sant, R.D.McConnell and A.Watts
Hydro Quebec Institute of Research, Canada

O. LJUNGSTROM, AERONAUTICAL RESEARCH INSTITUTE OF SWEDEN

Is it not true that the high power at high wind speeds now experienced should be regarded as a major breakthrough?

Authors' reply

Yes.

PAPER E3

The design, construction, testing and manufacturing of vertical axis wind turbines

R.Braasch
Sandia Laboratories, U.S.A.

A.J.ALEXANDER, LOUGHBOROUGH UNIVERSITY OF TECHNOLOGY, U.K.

The calculations were performed using a velocity power law exponent of 0.17. Can the author say how sensitive the answers are to variations in this exponent?

Author's reply

A plot, based on the same approach used in the paper, of energy cost versus machine diameter with wind velocity power law exponent as a parameter ranging from .10 to .24 is shown in Figure A1.

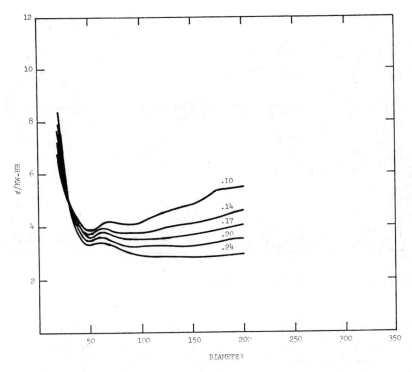

Fig. A1

O. LJUNGSTROM, AERONAUTICAL RESEARCH INSTITUTE OF SWEDEN

Congratulations on the interesting VA system family study, but should you not take a closer look at the blade upwind gale force wind parking case with the long, unsupported, slender blades? Perhaps introduce bicycle-type wire supports?

Author's reply

The loading of a parked, upwind blade has been investigated in a 150 mph wind (steady wind at 30 foot reference height, wind shear power law exponent of 0.17) with gravity. Several different blade end connections have been analyzed using MARC (a non-linear finite element computer code). Two configurations of interest are shown in Figures A2, A3 and A4. Figure A2 shows flatwise stress levels along the blade for a blade to tower connection called a mini-strut. The mini-strut allows a two position blade to tower connection where each connection is near the end of the blade, hence, the terminology mini. Figure A3 shows a comparison of the mini-strut configuration to a configuration with the strut portion removed. Figure A4 shows the deformed blade shape (amplified by a factor of 10) for the mini-strut design in the lower portion of the figure and for the unstrutted design in the upper portion of the figure.

The mini-strut design offers an inexpensive blade to tower connection approach. In addition, it is currently believed that tolerable stress levels will be achievable as suggested for the severe case as shown in Figures A2, A3 and A4.

We have been reluctant to consider wire-type supports because our analyses indicate that the high drag coefficient of these devices significantly degrade the rotor's aerodynamic performance.

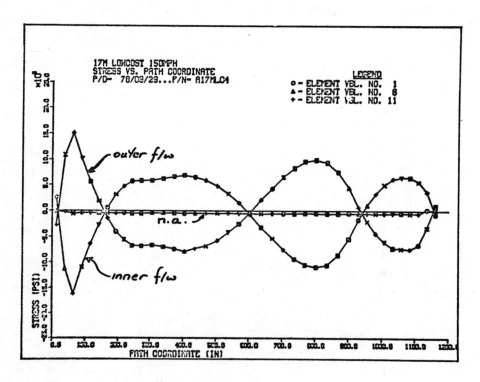

Fig. A2

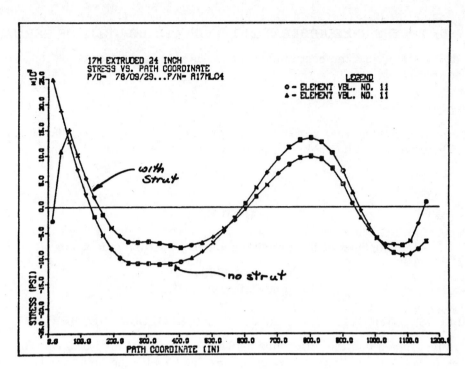

Fig. A3

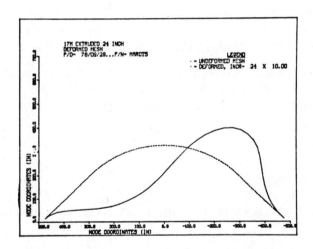

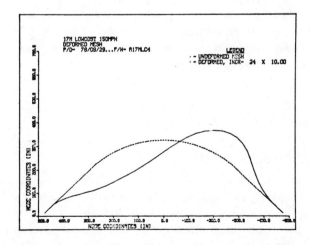

Fig. A4

P.H.THEYSE, THEYSE ENERGIEBERATUNG, FEDERAL REPUBLIC OF GERMANY

It was stated that the cost studies were based on a production of 100 machines. Produced at what rate?

Author's reply

The cost studies were performed for production levels of 10, 20, 50 and 100 MW_e per year. The results will be published in approximately April 1979.

PAPER E4

Development of the variable geometry vertical axis windmill

P.J.Musgrove and I.D.Mays
Reading University, U.K.

P.T.SMULDERS, EINDHOVEN UNIVERSITY OF TECHNOLOGY, NETHERLANDS

Do you agree that the difference between your way of measuring power output performance in the field and that of, for example, Sandia with the "bin" method could make a direct comparison of test results of your rotor and (Sandia's) Darrieus rotor questionable?

Authors' reply

Reply not received at time of publication

A.VOLLAN, DORNIER-SYSTEM GmbH, FEDERAL REPUBLIC OF GERMANY

Due to the periodically varying aerodynamic forces, the blades will oscillate about their hinge which will induce coriolis forces and torque moments in the radial beam.

Can you tell whether the resulting oscillating stresses are serious to the fatigue life of the rotor !

Authors' reply

Reply not received at time of publication

R.K.HACK, NATIONAL AEROSPACE LABORATORY NLR, NETHERLANDS

1. Are the bending stresses in the blades not going to be a serious limitation for large scale turbines of this concept?

2. Is it possible to construct the blades of large turbines as a pendulum, and without the spring construction? The speed control could be given by the gravitation forces, acting as spring forces. Have you also investivated the low frequency oscillations or even instabilities which may occur in this type of speed control?

Authors' reply

Reply not received at time of publication.

B.R.CLAYTON, UNIVERSITY COLLEGE LONDON, U.K.

1. It is characteristics of vertical-axis rotors that the incidence angle varies from a maximum positive to a maximum negative value each revolution of the rotor. Subsequent analysis to predict the aerodynamic performance of vertical-axis rotors seems to depend on lift and drag data for aerofoil sections mounted in a steady flow stream. Bearing in mind that the fluctuations of incidence angle for the authors 3-m diameter rotor is over 2Hz would the author expect that the mean C_D and C_L values for the model rotor are adequately represented

by the steady values at corresponding incidence angles.

     2. The question of unsteady, non-uniform flow takes on even greater inportance for that blade passing through the wake(s) shed by the upstream blade(s). Would the author kindly comment on this feature.

Authors' reply

     Reply not received at time of publication.

B.R.CLAYTON, UNIVERSITY COLLEGE LONDON, U.K.

     The author has given some details of his experiments on a high-solidity vertical-axis rotor with the increase in solidity being partly obtained by increasing the chord length of the blade. However, the ratio c/r_0 should not be made too large otherwise the effective flow incidence angle varies significantly along the chord. The assessment of C_L and C_D then becomes rather difficult and may be likened, for example, to the side force on a ship engaged in a tight circle maneouvre. Would the author please comment on this aspect in connection with the rotor illustrated and his further work on high solidity rotors. There is, of course, an advantage of low-aspect-ratio blades in that stalling incidence angle is much larger than that for a high-aspect-ratio blade with no loss of maximum C_L. Consequently, lower rotational speeds of the turbine can be tolerated, applicable therefore to direct pumping applications. Is this the philosophy behind the author's work?

Authors' reply

     Reply not received at time of publication.

PAPER E6

Wind tunnel corrections for Savonius rotors

A.J.Alexander
Loughborough University of Technology, U.K.

B.R.CLAYTON, UNIVERSITY COLLEGE LONDON, U.K.

     In 1977 experiments were conducted on a Savonius rotor at University College London with the principal aim of investigating the flow behaviour in and around the rotor. There was no intention of measuring the power coefficient since the small size of the model would have put any such measurements in considerable doubt. Instead, a flow-visualization programme was carried out with water as the fluid medium and using the hydrogen bubble technique which is now well developed in my department (Ref. 1). The rotor, made of perspex and painted matt black, was mounted horizontally in a water channel. The rotor had circular plan end walls and the complete rotor was supported by end walls which housed the bearings. The test section was bounded by the glass bottom of the channel and a perspex roof. The blockage coefficient, as defined by the author, was 0.33 and is therefore relevant to the author's work.

     Thin wire electrodes for generating the tiny hydrogen bubbles were embedded in the surfaces of each half of the rotor and across a diameter joining the rotor tips. In addition, a moveable wire electrode could be traversed within both upstream and downstream flow regions. Rotors consisted of a continuous S form along with several gap sizes between the two semicircular halves of the S. Cine films were taken of the illuminated bubbles as the electrodes were switched on in turn or in various combinations. Still photographs were also taken with the aid of a stroboscope but although these (and views with the naked eye) clearly portrayed the flow regimes reproduction in print probably would not. Thus, Fig. 1 is a sketch of a typical observation and it was perfectly evident that the strength of vortex K, was always much greater than K_2 in the position shown. Further more, the velocity V_c past the top ($+90^0$) of the rotor was considerably less than that past the bottom (-90^0). Moreover, the magnitudes of V_c were time dependent since they varied with the orientation of the rotor.

Under these circumstances it would seem that previous data based on the blockage of flat plates or discs are inappropriate for comparison with the blockage effect of Savonius rotors. Could it be that the flow behaviour described above accounts, at least in part, for the deviation from the flat plate shown in the author's Fig. 7?

Incidentally, in Fig. 7 is V kept constant? if so then presumably no account is taken of Reynolds number effects nor of the rate at which V_C fluctuates with rotational speed of the rotor.

Reference

1.   CLAYTON, B.R. and MASSEY, B.S. Flow visualization in water: a review of techniques. J.Sci.Instrum, Vol. 44, pp 2-11, 1967.

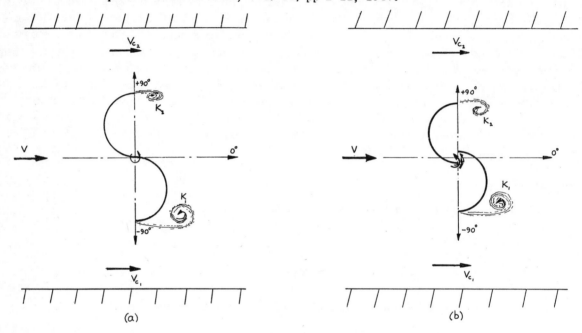

(a)   (b)

Fig. 1

Author's reply

The work carried out at University College on the flow patterns through Savonius Rotors is very interesting and should lead to a better understanding of the flow mechanisms.

Dr. Clayton says that the blockage is rather high, 33%, but this should not effect the flow pattern through the rotor to any great extent.

In his Figure 1 I do no find it surprising that Vc_2 is less than Vc_1. With the rotor turning anti-clockwise it presumably generates an anti-clockwise circulation which will produce a negative (downward) lift force in the tunnel airstream. This would require higher pressures and lower velocities at the top and vice-versa at the bottom, i.e. $Vc_2 < Vc_1$.

Since the vortex K_1 is stronger than K_2 however, this means there is a net anti-clockwise circulation in the flow which would seem to violate Kelvin's theorem.

Two sets of tests were carried out to form the basis of my paper. The first set merely repeated and extended Maskell's work on flat plates. The second set of tests used the same basic method but used Savonius Rotors. Admittedly the drag force measured was rather unsteady but it represented an average of all the conditions met within one rotation of the rotor.

The final cofirmation of the validity of the correction was obtained from the comparative tests carried out in the two tunnels at Cranfield and Loughborough. The same set of models was tested at the same wind speed but in an airflow of different size, so that only S/C was varied. Since the flow patterns round flat plates and rotors is so different it is not surprising that the corrections are rather different.

The airspeed was kept constant and therefore there could be some Reynolds number effect in the tests, but had it been appreciable it would surely have shown up in the comparative series of tests in the two wind tunnels.

W.L.SCHMIDT, WORLDWIND, CANADA

Would you care to comment on the discrepancy between your data and that from Sandia Labs?

Author's reply

I could not comment on the differences between the two sets of results without knowing what corrections had been applied to the Sandia data.

M.SERT, TUBITAK–MARMARA RESEARCH INSTITUTE, TURKEY

I would like to remark that instead of testing "a fairly large size Savonius Rotor" in a wind tunnel and trying to find out about the tunnel blockage corrections, it would be much better to test the model in the field, under realistic conditions, in order to obtain results of practical significance on such a specifically applied research subject.

Author's reply

One reason for carrying out the tests was to enable reasonable results to be obtained from small wind tunnels. In order to measure power output accurately the model windmill must usually be large compared with the tunnel. In such cases accurate corrections are essential.

We are proposing to test a Savonius Rotor in the field but in such a case it will be difficult to measure upstream conditions accurately owing to the vertical velocity gradient and large turbulence. The wind tunnel test, although idealised, does give an accurate and repeatable result with which outside tests can be compared.

PAPER E7

The cycloturbine and its potential for broad applications

H.Meijer Drees
Pinson Energy Corporation, U.S.A.

P.MUSGROVE, READING UNIVERSITY, U.K.

Mr. Drees referred to the fact that Darrieus, in his basic patent published in 1927, describes the cyclic pitch vertical axis turbine. Can I put in a word for Mr. Schneider, who almost concurrently published his patent for a vertical axis, cyclic pitch, turbine, designed for use in water, but basically identical to the cyclic pitch wind turbines we see today. This concept was taken up commercially and has led to the very successful range of Voith-Schneider propellers.

Could Mr. Drees indicate the typical amplitude of his cyclic pitch modulation, and would he also confirm that the quoted value of $C_P = 0.4$ is relative to $\frac{1}{2} \rho V^3 A$, not $\frac{16}{27} \times \frac{1}{2} \rho V^3 A$?

Author's reply

Reply not received at time of publication.

VERTICAL AXIS WINDMILL DEVELOPMENT BY BRITISH AEROSPACE

A.C.Willmer

British Aerospace, Aircraft Group, Bristol

The Aircraft Group of British Aerospace at Bristol is currently involved in the development of the variable geometry vertical axis windmill originated at Reading University and described by Musgrove (1). The present work is funded by the Department of Energy and is undertaken with Taylor-Woodrow and Reading University. It consists of a design study for a 25m diameter windmill for electricity generation. This is seen as a necessary intermediate step towards multi-megawatt sized windmills.

As part of this design study a series of tests have been completed in the British Aerospace wind tunnel at Bristol. The model is 3m diameter and has two blades 2m x 0.15m of NACA 0015 section. Tests were conducted in the wind tunnel return section, which has a cross section of 9.2m x 7.9m. The tunnel blockage area ratio is thus kept to 0.1.

To control model speed on the unstable parts of the characteristics and to start the windmill a drive/brake unit was provided. An electric motor is coupled to the windmill shaft via an eddy current variable speed drive and a second eddy current coupling is used as a brake. An automatic control system acts on these eddy current units to ensure close control of rotational speed. The complete drive/brake unit is mounted at the bottom of the vertical shaft below a false floor in the wind tunnel.

A torque meter and speed sensor are mounted on the shaft above the drive/brake unit. Strain gauges are fitted at various points on the model to measure instantaneous loads.

Wind tunnel corrections have been derived by adapting Glauerts method for propellors (2). The resulting correction agrees with the usual streamline body correction for low drag conditions and is close to Maskell's bluff body correction at high drag (Figure 1).

Performance predictions for the model are based on a single stream tube calculation which uses aerofoil data measured on one of the windmill blades mounted on the tunnel force balance. The calculated level of performance is confirmed by the initial test results (Figure 2) but the tip speed ratio for best performance is not well predicted.

To explain this difference between measurement and calculation, one must postulate that the blade section stall angle is effectively higher for the fluctuating flow conditions encountered on the windmill than for the blade alone in steady flow.

References

1.   Musgrove, P.J. "The Variable Geometry Vertical Axis Windmill" Proc. BHRA Wind Energy Symposium, Cambridge, Sept 7-9, 1976.

2.   Glauert, H., "The elements of aerofoil and airscrew theory" Cambridge University Press, 1948.

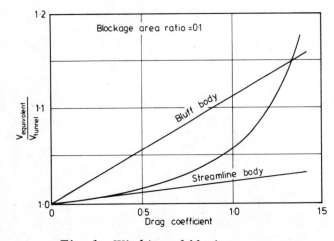

Fig. 1   Wind tunnel blockage correction

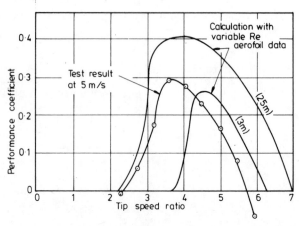

Fig. 2   Windmill performance

A NEW VERTICAL-AXIS WIND TURBINE
(Summary of film presentation)

L. Arnbak
Scangear Ltd, Denmark

A short presentation of a new wind turbine concept was given in the form of a film with comments. The vertical rotor is related to the rotating circular cylinder considered by Magnus, Prandtl and Flettner, but has a quasi-elliptical cross-section yielding superior torque and self-starting capabilities. The details of the cross-section were derived from the classical aerofoils and are shown in Fig. 1.

Major parameters are:

Rotor length: 600 cm
Major axis of cross-section: 132 cm
Minor axis of cross-section: 62 cm
Swept area: 8 m^2
Height overall, including pedestal: 10 m
Turbine power coefficient: approx. 0.1 (same range as Savonius rotor)
Diameter of streamline curl around rotor: some 25 m
Site: near North-Sea coast of Jutland, Denmark. (see Fig. 2.)

Performance data: Operating since erection in Nov. 1977 without malfunctions or mechanical problems, achieving to date some 20 million revolutions including full-speed operation during several winter gales. The very rugged construction in standard building elements does not impair the low-noise performance and self-starting from 4-5 m/sec wind speed.

Future plans: Model tests indicate increased power output from attaching guiding vanes.

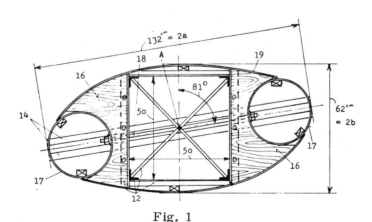

Fig. 1

Fig. 2

EXPERIMENTAL INVESTIGATION ON THE PERFORMANCE
OF A DARRIEUS-ROTOR IN NATURAL WIND

F. Rasmussen and B. Maribo Pedersen
Technical University, Denmark

In the year of 1977 a number of investigations were made on a 4m experimental Darrieus-rotor constructed at Department of Fluid Mechanics and erected at the site of Risø National Laboratories.

Rotor and control system

The rotor has the following data:

diameter:	4 m
diameter-height-ratio	1.0
wing profiles:	NACA 0012 with chord 0.25m
	or NACA 0018 with chord 0.15m
number of wings:	2 or 3

The wings were made of laminated wood that was bent into troposkien-form. Afterwards the wings were profilated and covered with a thin layer of glassfibre reinforced plastic.

On the basis of a DC-generator an electronic control system was constructed either to keep constant speed or constant velocity ratio on the rotor (Fig. 1).

The speed control is reached by controlling the magnetizing current to the generator, and thereby controlling and torque from the generator. The control system is a proportional-integral low voltage control system and with this system it was possible to keep constant rpm within one per cent.

In the case of constant rpm the reference voltage is a constant DC-voltage and in the case of constant velocity-ratio the reference voltage is a wind-speed proportional voltage.

Measurements

The measurements were performed using the method of bins, and a sampling frequency of one per second. Even at this high sampling frequency there was a good correlation between the wind velocity and the torque averaged over one revolution. Each power curve at constant rpm (Fig. 2 and 3) represents a data-collection for 10 minutes.

The measurements revealed a solidity-dependent behaviour of the power-curves in the stalling region. At high solidities the power-curves were increasing with increasing wind speeds, and at lower solidities the power-curves showed a decreasing tendency. In the last case the result was a expected from the "multiple stream tube model".

The corresponding Cp-curves (Fig. 4) show a strong dependency on the Reynolds number. The maximum efficiency occurred at lower tip-speed ratio than expected, and 0.4 was exceeded at a higher solidity.

The measurements with constant tip-speed ratio were combined to statistical wind velocity distributions and showed an increase in mean energy output of 30 per cent as compared to the optimal mean energy output at constant rpm.

According to noise it was discovered that the noise emission from the NACA 0018 profile was somewhat higher than that for the NACA 0012 profile.

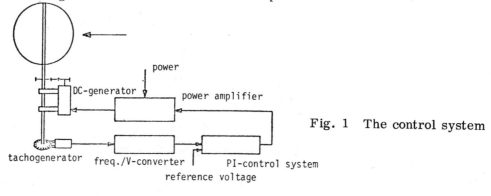

Fig. 1   The control system

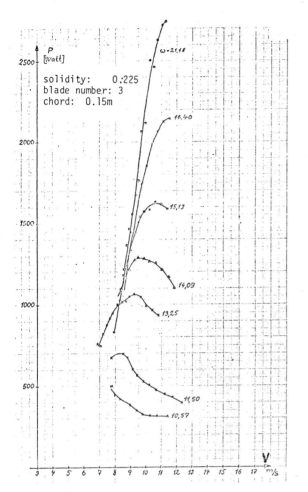

Fig. 2   Power curves

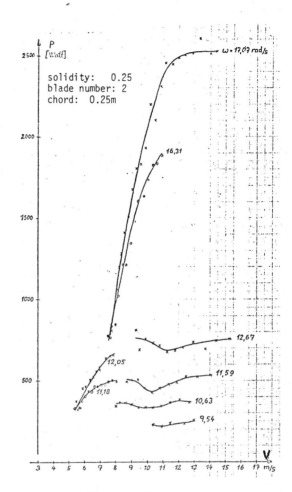

Fig. 3   Power curves

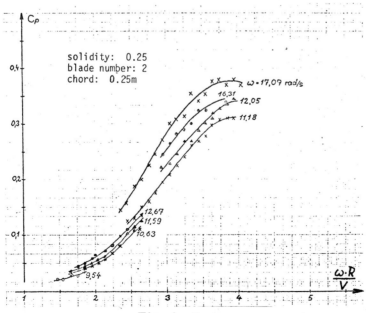

Fig. 4   Cp-curves

CONTRIBUTION EX4

PRACTICAL CONSIDERATIONS IN THE DESIGN
OF A VERTICAL-AXIS WINDMILL

F.C.Evans
University of St Andrews, U.K.

The design procedure for a vertical-axis prototype windmill is discussed, with particular relation to the use of pivotted blades, the position of the supporting arms and the proportions of the machine. The resulting design is shown to be self-starting and to be capable of exploiting the high available power at high wind speeds. Results from two prototypes, one of 6 m^2 swept area, and one of 24 m^2 will be reported at the Symposium.

This paper gives an account of the thinking (some of it beforehand and some hindsight) which has accompanied the production of a range of vertical-axis windmills at the University of St Andrews, since 1975.

1. Design requirements

Many modern designs of windmill operate with efficiencies in the range 20% to 40%, and this is significant fraction of the maximum attainable efficiency, which is 59.3% on the simple theory, and probably not very different from that Figure on any more sophisticated theory. It therefore seems to the writer that extreme effort directed at raising the maximum efficiency of a particular design by a few percent is not usually worthwhile, although of course attention should be paid to routine points of design such as minimising the drag of all moving parts of the structure.

The points regarded as important in this design are
(i)   robustness
(ii) self-starting ability
(iii) a fairly flat curve of efficiency versus wind speed.

The general point about robustness is illustrated by the fact that at least one large-scale wind generator system is known to have failed catastrophically in each of the USA and the UK and this represents a large waste of capital expenditure. More fundamental, however, is the point that all windmill designs have to have some speed-limiting mechanism to prevent damage by centrifugal force in high winds, and if one can raise the speed at which this mechanism begins to operate by a small amount, a large gain in average annual energy conversion can be produced, in suitable sites.

Self-starting is a more important attribute than it may seem at first sight. The two most common starting mechanisms proposed for inherently non-self-starting machines are an auxiliary windmill and an electric starter motor operated by a battery. The auxiliary windmill can cause a large increase in rotational drag which will reduce efficiency, and in backthrust which will increase the overturning moment. The energy used by an electric starter is won at great expense; when the respective efficiencies of the windmill, the gearing to the alternator, the alternator, the rectifier, the chemical process of battery charging, the process of discharging, the starter motor and the gearing to the windmill are all multiplied together it is not surprising that the few thousand joules required to start the windmill may represent many thousands of joules of original wind energy. Studies of wind speed distributions with time indicate that this starting process may have to be performed about 20 times a day.

The ability to extract a reasonable proportion of power over a wide range of wind speeds may be important according to the circumstances. If the main quantity of interest is the total annual energy conversion, then there is little to be lost by using a windmill which is inefficient or inoperative at low wind speeds, because of the cube law relating power and wind-speed. However, a common situation might be that the bulk of the energy is stored in a heat reservoir in the form of a tank of hot liquid and used to feed a central heating system. It then becomes important to be able to keep the circulating pump of the central heating system going for as large a fraction of the time as possible, in order to cut down on the size of storage batteries which represent a high initial capital outlay with a high rate of depreciation. In this case, the small amount of power available at low wind speeds becomes disproportionately important. The same is true where a small amount of power is required to operate lights, commercial equipment such as calculators and cash registers, and domestic equipment such as radio and television.

2. Effect of pivotted blades

All the machines built and designed at St Andrews, so far, have had each blade pivotted about a vertical axis at about $0.33c$ aft of the leading edge, where c is the chord of the blade. The blades are statically balanced about this axis, which is approximately equivalent to balancing the centrifugal forces about the same axis provided the pitch angle is small. The pitch angle is determined, in the steady state at least, by equilibrium between the couples due to the aerodynamic force and a control spring. This arrangement is shown diagrammatically in Figure 1. If the spring is weak, the blade tends to reach equilibrium at an angle of attack just below the stall angle which is close to the value for maximum lift. If the blade tends to stall, the centre of pressure moves aft towards the $0.5c$ line and the resulting couple decreases the angle of attack. If the angle of attack is too small, the centre of pressure moves forwards towards $0.25c$ and the angle of attack increases. The effect of the spring is to bias the equilibrium in the direction of decreased angle of attack, according to its strength.

The self-acting pivotted blades give the machines their properties of self-starting and of useful power production over a wide range of wind speeds. They also avoid the very high stresses associated with variable-pitch mechanisms using cams and rods. At very high rotational speeds (of the order of 5 rev sec^{-1}) it is likely that the natural resonant frequency of the blades will be less than the rotational frequency, so the amplitude of pitch variation will become very small. The operation of the machine is then virtually identical with that of a machine of fixed zero pitch.

The pivotted blades can also be used to incorporate the speed-limiting effect in the design of the machine. If the C.G. of each blade is very slightly aft of the pivot axis then the blades will tend to swing outwards at high rotational speed, providing an air-brake effect. Adjustment of the balance of the entire blade in this way is rather a delicate procedure, and it may be better to use a small portion of the blade area, i.e. a centrifugally controlled flap.

A further advantage of pivotted blades is indicated by the computer studies of Sarre (Ref. 1) which show that solidities above the usual values of around 0.1 are not a disadvantage in variable-pitch machines, although they are a disadvantage in fixed-pitch machines. This means that a single pattern of blade can be used with a range of machines of different diameters or different numbers of blades.

3. Vertical position of supporting arms

If a blade with uniform load w per unit length and length l is supported by a single support at the centre or by two supports at the ends the maximum bending moment is $wl^2/8$ and this is the factor which determines the speed at which the windmill must be governed. The predominant component of w is usually the centrifugal force mv^2/r where m is the mass per unit length, v the linear velocity and r the radius, since this is usually much larger than the aerodynamic force.

In order to minimise the maximum bending moment, for a blade of uniform section, the supports should be positioned at $l/2+2\sqrt{2}$ from each end, i.e. about $0.207l$ from the ends and this reduces the maximum bending moment to about $wl^2/46.6$, i.e. by a factor of $1/5.8$. This means that the rotational speed can be increased by $\sqrt{5.8}$ i.e. 2.4 times before the governing action begins, and assuming that the windmill speed is proportional to wind speed, this corresponds to an increase of the available power at limiting speed by a factor of 14. Provided the windmill is situated at a suitably windy site, this will give a significant gain in average annual energy conversion. There is a penalty in the form of increased drag caused by using two supporting arms instead of one and this will undoubtedly cause a decrease of a few percent in the efficiency. For this reason, it would probably be better to use a single supporting arm in sites of very low average wind speed and accept a lower governing speed. It is also possible that three or more supports, at the optimum positions, might be desirable in extremely windy sites, but this is not likely to be the case in the inhabited regions of the world. The machine described in section 5 is designed to be governed at wind-speeds of 16 m s^{-1} and upwards and to produce some useful power at these wind speeds, i.e. it does not shut down, and there is little to be gained by increasing this speed if the increased drag is taken into account.

4. Proportions of the windmill

Consider a machine of blade length ℓ and diameter d such that the swept area ℓ d is held constant. The ratio   max. bending moment/stiffness   of the blades increases rapidly with ℓ, being proportional to ℓ^3 if the blades are assumed to be monocoque constructions of thickness proportional to d, or to ℓ^4 if the blades are of constant thickness. From the point of view of strength of the blades it is therefore desirable to make the ratio ℓ/d as small as possible. The limit is set by the aspect ratio of the blades, the number of blades and the solidity of the machine. Taking an aspect ratio of 6 as the minimum desirable, three blades in order to give reasonably uniform torque, and a solidity of 0.3, the limiting ratio is ℓ/d = 0.6 and most workers have chosen proportions close to this figure. Again, a drag penalty has been consciously accepted by using three blades instead of two or one, for the sake of almost constant torque.

5. A prototype design

A prototype machine of dimensions ℓ = 2m, d = 3m has been designed and built. The supporting structure is also designed to support a machine of dimensions ℓ = 4m, d = 6m, and it is planned to build this later.

The blades have the outermost 3ℓ/16 of their length tapered to half the maximum chord, with the remainder of the length parallel-sided, in such a way that the 0.33c line remains straight. Corners have then been radiussed to give a shape which is a compromise between good aerodynamic performance and simplicity. The effect of the radiussed corners has been ignored in the calculations. The maximum chord is 333.3mm and the aspect ratio approximately 6.6. The aerofoil section is NACA 0015. The radius of gyration of a thin-walled shell of this shape has been determined by a graphical method as 15.9mm. Numerical values of the second moment of area for various shell thicknesses have also been abtained graphically. The blades are made of g.r.p. using unindirectional woven roving for maximum tensile strength along the length, laid up to 4mm thickness with polyester resin.

With the plan form described above, the optimum position for the supports is modified slightly, compared with the value mentioned in section 3, to 0.23ℓ with a maximum bending moment of wℓ^2/54.4. The supports are also positions of maximum bending moment, so it is not a good idea to make holes in the g.r.p. shell at these points for the supporting arms, and the latter have been offset internally by means of a steel spar, so that the arms can emerge at points close to the points of zero bending moment, which with this shape are calculated to be at 0.31 from each end of the blade.

The rotational speed is governed at 300 rpm which corresponds to a linear speed of 47 m s^{-1} and an acceleration of 148g. This allows full efficiency, at tip-speed ratio of 3, at wind-speeds up to 16ms^{-1} and reduced efficiency thereafter. With g.r.p. structures the stress is limited by strain rather than by tensile strength, i.e. excessive bending will occur before breakage, but even bearing this in mind there is a large safety margin in the design and the calculated minimum radius of curvature due to bending of the blades is 50m. Sketches of the construction are shown in Figure. 2.

The supporting mast is a stationary steel tube with a rotating steel tube inside it. The bending moment on the inner steel tube at the upper ball race is minimised by offsetting the sets of supporting arms equal distances above and below the ball race. The outer sections of the supporting arms are covered by grp fairings in order to reduce drag, with flexible rubber fillets to avoid the presence of sharp corners between the blades and the arms.

The load is a large-diameter alternator from the range of such machines designed by Mr H Winterbotham to generate several KW at 300 rpm. It is directly coupled to the windmill shaft, so that there is no gearing loss, and the design is such that copper and iron losses are very small.

6. Acknowledgements

I am indebted to Mr T Winnington, University of East Anglia for discussion of the effect of movements of the centre of pressure, to Dr R Thoms of the University of Cambridge for information on the number of starts per day likely to be needed, and to Mr H Winterbotham for many helpful technical discussions. The work has been privately financed and carried out by me in what spare time remains from my other research and teaching duties, but I am very grateful to the University of St Andrews for encouragement, use of workshop facilities, provision of a site and assistance with such matters as Planning Permission.

7. References

1.   P.E.Sarre   "Wind Energy Project" Report by Dept of Chemical Engineering, University of Exeter.

2.   P.J.Musgrove   "The Variable Geometry Vertical Axis Windmill", Int. Symposium on Wind Energy Systems, 1976, paper C7

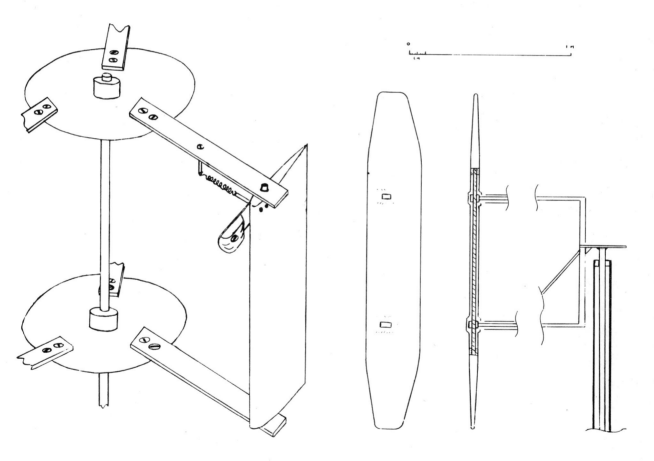

Fig. 1   Fig. 2

Discussion & Contributions

SESSION F: DESIGN AND CONSTRUCTION: CONCENTRATORS

Chairman: P.B.S. Lissaman, AeroVironment Inc., U.S.A.

Papers:

F1 Design and construction of a pilot plant for a shrouded wind turbine.
O. Igra and K. Schulgasser, Ben Gurion University of the Negev, Israel.

F2 Tipvane research at the Delft University of Technology.
Th. van Holten, Delft University of Technology, Netherlands.

Contributions:

FX1 Vortex augmented wind energy conversion.
P.M. Sforza, Polytechnic Institute of New York, U.S.A.

Note:
The papers were presented by the Authors whose names appear in bold print.

DISCUSSION

PAPER F1

Design and construction of a pilot plant for a shrouded wind turbine.

O. Igra and K. Schulgasser

Ben Gurion University of the Negev, Israel

D.J. MILBORROW, CENTRAL ELECTRICITY GENERATING BOARD, U.K.

The authors are to be congratulated on their clear presentation; it would be helpful, however, if the authors could quote conventional values of performance coefficient so as to give a rough estimate of the overall effectiveness of these novel techniques.

Authors' reply

Conventional bare wind turbines have efficiencies of not more than 60%. The proposed axial flow turbine has an efficiency between 85% to 95% (see O. Igra, Energy Conversion, **15**, pp. 143-151, (1976)). One should note that this improvement is in addition to the power augmentation resulting from the shroud.

D. LE GOURIERES, DAKAR UNIVERSITY, SENEGAL

Dr. Igra has said to us that the power developed by the wind turbine with diffuser was three times higher than the power developed by a wind turbine of the same diameter but without diffuser placed in the same wind. I think it is interesting to compare the power developed by the turbine with diffuser not to a turbine of same diameter but to an ordinary turbine of a diameter equal to the maximum diameter of the diffuser. What is the result in this case? Can Dr. Igra give an answer to the question? The problem is the problem of utility of diffuser.

Authors' reply

If we make the comparison in the form posed by the questioner then the redefined power ratio will of course be smaller or at most equal to unity. However as long as this ratio is close to one the suggested aerogenerator is much better than a bare propellor with diameter equal to the largest shroud diameter since for the shrouded type we obtain almost the full available power. Present machines, using the most sophisticated technology, do not generate even 60% of the theoretical limit. But the real crux of the proposed shroud is the contention that for large power output units, the replacement of very large rotating elements by very large stationary elements sufficiently reduces the technology and complexity so as to result in a significant overall cost reduction.

PAPER F2

Tipvane research at the Delft University of Technology.

Th. van Holten

Delft University of Technology, Netherlands

D. LE GOURIERES, DAKAR UNIVERSITY, SENEGAL

Mr. Van Holten has given no results concerning the increase of the power developed by the turbine with tipvane. I shall be glad to know what is the relative increase of the power

power compared to a wind turbine of the same diameter but without tipvanes and to an ordinary wind turbine whose diameter is the sum of the diameter of the turbine with tipvane plus the total length of the tipvanes.

Author's reply

For detailed performance estimates I may refer to the paper E3 presented at the first symposium in Cambridge.

Roughly the situation can be summarised as follows:

We can augment the mass flow by a factor of 4, which means that the gross power augmentation will be around 4. Conservative estimates regarding the additional losses due to frictional drag of the tipvanes give us a net power augmentation by a factor around 2. This doubling of power output is relative to a conventional turbine of the same diameter. Compared with a conventional turbine having the same total blade area, the gain is of course less and is estimated to be in the vicinity of a factor 1.5.

It should be emphasised that according to these estimates a tipvane turbine still has a power advantage when compared with a larger conventional turbine where the blade area is equal to the area of our turbine blade increased with the tipvane area.

The latter is the cause of the improved specific costs indicated by the studies of Aero-Vironment.

D.J. MILBORROW, CENTRAL ELECTRICITY GENERATING BOARD, U.K.

The authors are to be congratulated on their clear presentation; it would be helpful, however, if the authors could quote conventional values of performance coefficient so as to give a rough estimate of the overall effectiveness of these novel techniques.

Author's reply

The above mentioned doubling of power output would lead to power coefficients of the order of 0.9. This is higher than the conventional Betz limit, the reason being that the affected stream tube is larger than according to Betz's theory.

Consequently the factor by which the power is non-dimensionalised in Betz's theory is really not relevant for the tipvane turbine.

B. SØRENSEN, NIELS BOHR INSTITUTE, DENMARK

Based on an energy per weight comparison, it would seem, that the gain due to the augmentation factor for the tipvane machine is offset by the increase in material use. Also, the augmentation factor is proportional to the length of the tipvane, if I remember correctly, so there would seem to be little prospect for reducing the length of the tipvanes. Could you comment on the dimensioning of the vanes?

Author's reply

For the first part of the question I may refer to the above. Secondly the remark that the tipvane span is a given quantity: this is correct as long as the tipspeed ratio is kept constant.

In fact the span is determined by the condition that the rotor runs at synchronous or over-synchronous speeds. However, the vane area can nevertheless be reduced when the lift coefficient is increased. This allows a reduction in chord length. Evidently it is very important to develop suitable aerofoils, capable of operating in curved flow at a high maximum lift coefficient, but still keeping the drag coefficient sufficiently low.

I. CARLLSON, LUTAB, SWEDEN

Does a tipvane turbine have to run at a constant tipspeed ratio?

Author's reply

There is no need to keep the rotor at a constant tipspeed ratio, as long as the speed surpasses the synchronous speed. This particular tipspeed ratio is called the synchronous ratio because in this case the tipvortex from the upstream vane tip exactly hits the downstream tip of the next vane, so that this is the lowest speed at which complete obliteration of the vortex exists. Below it there is almost no mass augmentation. Above the synchronous speed the tipspeed ratio may be varied without consequences for either mass flow or induced drag.

B.R. CLAYTON, UNIVERSITY COLLEGE, LONDON, U.K.

The tip augmentor described by the author refers to the operation of horizontal axis rotors containing two blades.
1)   Does the author expect that by proper selection of the tip-speed ratio similar augmentation could be obtained from a single-bladed or multi-bladed rotor?
2)   Does the author anticipate that augmentation using tipvanes could be expected if these vanes were attached to the tips of the blades of a vertical axis rotor of the type described in paper E4.
If not, could he suggest any other form of augmentation which may be applicable to vertical axis rotors?

Author's reply

1)   In principle it is possible to use tipvanes on single bladed rotors. However, one can expect that the necessary span for vortex synchronization becomes very large, and perhaps unrealistically large. It would of course depend on the design tip speed ratio.
2)   The principle of mass agumentation depends on the generation of cross-wind forces. If such forces are present, then always a venturi effect may be expected. Therefore vanes attached to the tips of a Musgrove rotor could have an augmentation effect, although it is not known whether in this case the balance between gross augmentation and parasitic losses would be favourable. Another system to generate crosswind forces could be the proper pitch variation of straight bladed Darrieus type rotors (so-called cyclo-gyro layout).

O. LJUNGSTROM, AERONAUTICAL RESEARCH INSTITUTE OF SWEDEN

How do you analyse 25% cost reduction for the tipvane rotor vs conventional without having the final answer on the T-bar type (or T-tail type) blade-tip dynamics problems, flutter, fatigue, etc.

Reply to question of O. Ljungstrom by:

P.C.Hensing, Delft University of Technology, Netherlands

A theoretical study is going on concerning the aero-elastic stability of a turbine with tipvanes. In a first approach the tipvanes are schematized to masses which are flexibly attached to the rotorblades. The aerodynamic forces acting on the rotor-blades are taken into account, whereas the aerodynamic consequences of the addition of the vanes are neglected, so that only the mechanical influence of the vanes is studied. The result of this first study is that under certain conditions the addition of the vanes might cause stability problems.

Further examination is necessary, such as the includion of aerodynamic forces on the vanes and the aerodynamic interference between vanes and rotor blades in order to get a rather complete view of the aero-elastic behaviour.

J. SHAPIRO, CIERVA ROTORCRAFT LTD., U.K.

Tip vane rotor with the same tip speed ratio will have higher rpm than a conventional rotor absorbing the same power. Consequently, the gear ratio would be lower and the gear, smaller. This cost saving should be considered.

O. LJUNGSTROM, AERONAUTICAL RESEARCH INSTITUTE, SWEDEN

1. Have you considered Hydro-propulsors, utilising the tipvane principle. Perhaps cavitation will be a major problem since flow field pressure minima are very strong?

2. Have you considered tipvanes for sea current turbines?

3. When you apply flow inducing propellers on vertical axis at top and bottom of Darrieus type VA/cross flow turbines, these I believe will have to run at much higher rpm than the Darrieus itself?

Author's reply

1. The application to thrusting rotors including ship-propellers has been considered very briefly and superficially. In fact some performance calculations for thrusting rotors have been published in the recent proceedings of the 12th ICAS Congress, paper No. B2-15.

Nothing is known yet about phenomena like cavitation etc.

Reply by P.B.S. Lissaman to points 2 and 3 not received at time of publication.

R.K.HACK, NATIONAL AEROSPACE LABORATORY NLR, NETHERLANDS

1) What's the effect of the rotor blade tip vortex on the tipvane tipvortex system and vice versa?

2) How did you separate the viscous drag and induced drag of the tipvane in your experiments?

Author's reply

1) The influence of turbine blades on the tipvane rotor system was simulated roughly during the model experiments, by choosing the flow blockage due to the vane struts such that this blockage equals the effect of real turbine blades.
The flow visualisations have shown that the effect is to give to the vane vortex a radial convection velocity. This is the reason why a certain amount of tilt of the vane is needed. The details of the flow field will be different when actual turbine blades are present, in particular because of the strong concentrated tip vortex from such a blade. Theoretical predictions are that an additional anti-symmetric spanwise lift distribution is superimposed on the existing lift distribution, whereas on the turbine blade the lift distribution changes in such a way that a finite amount of lift still exists at the blade tip. Experiments with a complete turbine - vane model have started recently but results cannot be given yet.

2) One of the ways to separate viscous and induced drag was described in the written paper, on page F2-17 of Volume 1.
Later another method was used, which also indicated that the induced drag is practically zero. This was done by measuring a drag polar of a tipvane and by applying a very simple form of regression analysis. Any induced drag is bound to give a parabolic content in the curve of total drag coefficient versus lift coefficient. The measured drag polars appeared to be almost perfectly linear with the lift coefficient which gives rise to the conclusion that no induced drag is present.

VORTEX AUGMENTED WIND ENERGY CONVERSION

P.M. Sforza

Polytechnic Institute of New York Aerodynamics Laboratories, U.S.A.

Summary

A discussion of research, design and development on a novel aerodynamic device which can concentrate and augment natural winds is presented.

THE VORTEX AUGMENTOR CONCEPT

Under certain predictable conditions vortices appear in a flowing fluid. The thrust of the present work is the utilization of the unusual aerodynamic characteristics of vortices to develop improved wind energy conversion systems.

The keystone here is the generation and control of discrete vortices of high power density by appropriate interaction of aerodynamic surfaces with natural winds of relatively low power density. Suitably designed turbines are used to extract energy from this compacted vortex field. This idea is termed the Vortex Augmentor Concept (VAC)*.

THE VORTEX FLOW FIELD

Vortex flows best suited to the VAC are those established by flow separation from highly swept sharp leading edges typical of supersonic aircraft planforms (Refs. 1, 2). Characteristics of the vortex field generated by delta-type aerodynamic surfaces are fairly well understood and have received considerable attention (Ref. 3), but detailed measurements of the velocity field over these surfaces have received scant attention.

Experimental results for the vortex velocity components from the few studies available (Refs. 4 and 5) indicate that not only are the axial velocity components in the vortex appreciably larger than the free stream velocity but, in addition, there is a circumferential velocity component present whose magnitude is on the order of that of the free stream velocity. This then is the augmenting effect of the vortex generating surface: it amplifies the wind speed within the vortex field in that the swirling flow tends to concentrate the low energy flux wind from a large upstream area into a high energy flux flow in a small (vortex) area). For example, integration of the experimental data for energy flux shows an increase in available power of 2 to 9 times that in the undisturbed oncoming flow per unit area.

VAC WIND TUNNEL STUDIES

Laboratory scale VAC systems (Ref. 6) were the subject of over 100 wind tunnel tests. The effect of the vortex augmentation offered by a simple delta surface alone is to provide a three-to fourfold increase over unaugmented operation. Various configuration changes can boost the increase to a factor of about five. Detailed flow investigations are being carried out using a fully motorized, five degree of freedom probe traversing system in our 1.2 x 1.5 m wind tunnel. Results for candidate VAC planforms are given in Ref. 7.

ROTOR DESIGN AND TEST

Turbines for transforming the flow power to shaft power are subject to the whirling vortex flow described previously. Thus, the vortex augmentor surface acts so as to process the wind prior to its reaching the turbine. Nonuniformities in the wind profile will tend to be

---

* Patented and patents pending.

smoothed out during the process of vortex formation and propagation. When the small size of the rotors required by the VAC system is also considered it is evident that high rotational speeds are possible. Thus there can be reduced requirements for step-up transmissions to the generator.

A rotor test facility has been constructed to improve design capability; model rotors up to about 1.5 m in total span may be tested at speeds up to about 10m/sec. Instrumentation includes a dynamometer as well as a five degree of freedom traversing system, similar to that used in the wind tunnel tests, for detailed rotor flow field measurements.

PROTOTYPE DEVELOPMENT

After carrying out a great deal of basic research and experimentation on laboratory scale systems which validated the Vortex Augmentor Concept, a prototype for field testing has been developed (Ref. 8). A simple flat plate delta planform was chosen for the prototype augmentor surface. The power train for the prototype includes rotors, driveshafts, a torque-meter, and an electric brake. With this system it is possible to obtain detailed torque, rotational speed, and power measurements (Ref. 9). Directional control of the prototype is by aerodynamic means, i.e. a vertical rudder positioned along the centerline of the vortex augmentor.

The prototype is being tested in the field (Fig. 1) within a ring of wind direction and velocity sensors (Ref. 10). Data from the wind survey network and from sensors on the proto-type is processed by an on-line minicomputer. In addition, further studies relating to vortex flow field analysis and experiment, rotor system theory and experiment, stability and control, and structural analysis and design are being carried out. The information obtained is being utilized to formulate economic assessments of the Vortex Augmentor Concept.

ACKNOWLEDGEMENTS

Support for this research was due in part to the U.S. Energy Research and Development Administration (under contract no. E(49-18)-2358) and the Polytechnic Institute of New York.

REFERENCES

1. **Sforza, P.M.:** "Aircraft vortices-benign or baleful". Space/Aeronautics, pp. 42-48, April 1970.
2. **Thwaites, B.:** "Incompressible aerodynamics". Oxford, 1960.
3. **Parker, A.G.:** "Aerodynamic characteristics of slender wings with sharp leading edges - a review". Jl. of Aircraft, 13, 3, pp. 161-168, March 1976.
4. **Earnshaw, P.B.:** "An experimental investigation of the structure of a leading edge vortex". ARC R and M No. 3281, March 1961.
5. **Hummel, D.:** "Undersuchungen uber das aufplatzen der wirbel an schlanken deltaflugeln:," Zeitschrift fur Flugwissenshaft, 13, Heft 5, 1965.
6. **Sforza, P.M.:** "Vortex augmentor concepts for wind energy conversion". Wind Energy Conversion Systems, Second Workshop Proceedings, edited by F.R. Eldridge, MITRE Corp., MTR-6970, NSF RA-N-75-050, September, 1975.
7. **Sforza, P.M., et al:** "Flow measurements in leading edge vortices". AIAA 15th Aerospace Sciences Meeting, Paper No. 77-11, January 1977. Appeared in the AIAA Jl., 16, 3, pp. 218-224, March 1978.
8. **Sforza, P.M., et al:** "Vortex augmentors for wind energy conversion". Proceedings of the International Symposium on Wind Energy Conversion, Cambridge, England. British Hydromechanics Research Association, April, 1977.
9. **Sforza, P.M., et al:** "An experimental facility for wind engineering research". AIAA 10th Aerodynamics Testing Conference, Los Angeles, CA., April, 1978, Paper No. 78-813.
10. **Sforza, P.M. and Stasi, W.:** "Wind power distribution, control and conversion in vortex augmentors". Fluids Engineering in Advanced Energy Systems, C.H. Marston (ed)., ASME, N.Y., pp. 45-57, 1978.

Fig. 1

Discussion & Contributions

SESSION G: APPLICATIONS – LARGE SCALE

Chairman: Th. Ykema, KEMA, Netherlands

Papers:

G1   The regulation of an electricity supply system including wind energy generators.
B. Sørensen, Niels Bohr Institute, Denmark

G2   An economic model to establish the value of the WECS to a utility system.
E.E. Johanson and M.K.Goldenblatt, JBF Scientific Corporation, U.S.A.

G3   Implications of large scale introduction of power from large wind energy conversion systems into the existing electric power supply system in the Netherlands.
G.H. Bontius, A.H.E. Mander and Th. Stoop, NV KEMA, Netherlands

Notes:

The papers were presented by the authors whose names appear in bold print.

Paper G2 was presented by L.V. Divone, Department of Energy, U.S.A.

DISCUSSION

The regulation of an electricity supply system including wind energy generators

B. Sørensen

Neils Bohr Institute, Denmark

J. SHAPIRO, CIERVA ROTORCRAFT LTD., U.K.

I was always convinced that wind power for public networks should be treated as negative demand, until energy storage becomes available. However, I now learn the remarkable fact that wind-generated electrical energy cost is only 50% higher when you throw away half the wind energy as Professor Sørensen shows. We have witnessed in the last five years increases of 100-300% in energy cost yet our sensitivity curves only extend to margins of 20-30%. Economic options should be made available for ranges of energy cost fluctuations of the same order as those we have witnessed and survived. Energy storage systems (mainly thermal) exist in which the cost as a function of stored energy follows the type of curve given by Professor Sørensen as the justified investment into energy storage.

A high percentage of nuclear energy in a grid will require energy storage as much as a high percentage of alternative energy; otherwise nuclear energy will become much more expensive than its present forecasts because of reduced load factor. The development of cost-effective energy storage is therefore essential and should not be debited against alternative energy alone.

Author's Comments

The curves commented on by Mr. Shapiro were shown at the symposium but are not included in my paper. I therefore reproduce them here. A detailed discussion of the calculations upon which the figures are based may be found in Ref. 8 in my paper (in press).

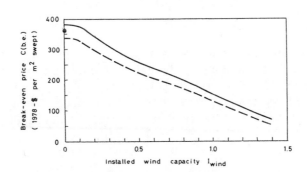

Fig. G1-6   Break-even price for systems of the type discussed in the paper. Solid line: Cost per start of back-up or call of peak-power unit is $ 2000. Dashed line: The cost is $ 10,000.

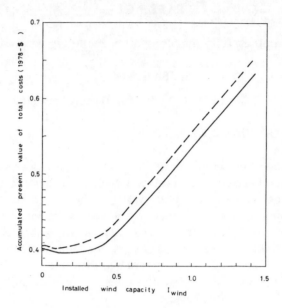

Fig. G1-7   Total cost of supplying 1 kWh per year in 25 years. Assumed wind converter cost: 300 (1978-) $ per m^2 swept by rotor.

R.MEGGLE, MESSERSCHMITT-BOLKOW-BLOHM GmbH, FEDERAL REPUBLIC OF GERMANY

In your simulation you applied wind prediction in order to decide on replacement of base-load units or intermediate-load units respectively.

The predicted wind power (to replace a baseload unit) is taken as the average over a preceeding 24 hour period and is applied over a following 24 hour period delayed by 6 hours. For intermediate-load an averaging over 4 hours is used.

What is the time-correlation of the wind data you used?

Author's reply

Sequential hourly data for a typical year were used in the simulation, and some insight in the time-correlations present in these data can be gained from the Fourier analysis shown as Fig. 5 in my paper. An alternative discussion of time-correlations in similar Danish wind data may be found in "Vindkraft i Elsystemet", Report from an ad hoc working group the "The Danish Academy of Technical Sciences" (Polyteknisk Forlag, Copenhagen 1977).

PAPER G2

An economic model to establish the value of the WECS to a utility system

E.E. Johanson and M.K. Goldenblatt

JBF Scientific Corporation, U.S.A.

P. MUSGROVE, READING UNIVERSITY, U.K.

Some of the slides you presented, showing the value of WECS, seemed to differ significantly from the figures given in the published paper. E.g. Fig. 13, on page G2-24, indicates a maximum value of $800/kW whereas the corresponding slide showed a value of

$1,000/kW. And Fig. 14, also on page G2-24, indicates a maximum value of $400/kW whereas the corresponding slide showed $800/kW. Can you confirm that the higher figures shown on your slides represent your most up-to-date assessment of the value of WECS?

Author's reply

Yes.

PAPER G3

Implications of large scale introduction of power from large wind energy conversion systems into the existing electric power supply in the Netherlands

G.H. Bontius, A.H.E. Mander and Th. Stoop

NV KEMA, Netherlands

P. MUSGROVE, READING UNIVERSITY, U.K.

You indicate that assimilating a significant energy input (e.g. 15% from wind energy systems into the Dutch grid would cause problems, because the peak wind energy output could then exceed the present grid capacity. This problem results from the fact that you have assumed a wind turbine design having a load factor of only 17%. (Top of page G3-29, the statement is made that the annual utilisation factor for a large WECS would be only 1500 h, i.e. 17%). Why did you choose such a low load factor? The load factor is a parameter that is under the control of the wind turbine designer, and it is worth noting that the JBF Scientific Corporation study (Paper G2) assumes the use of wind turbines with a 44% load factor. The use of wind turbines with a load factor of about 40% will surely greatly reduce the problem of integrating wind energy systems into the grid.

Authors' reply

With given wind characteristics on a given location, increasing the load factor (that is: reducing the generator capacity) means reduced overall energy production. As a consequence thereof, more WECS have to be installed in order to obtain the same amount of energy. After all the peak power output would remain unchanged, roughly speaking.

It might, however, well be that our estimated utilisation time of 1500 h (for a given location and a more or less arbitrarily chosen maximum power output of 1 MW) is somewhat too pessimistic. We do not know and nobody seems to know. One of the most important questions remains unsolved for the time being, namely, what is a fair and realistic estimation of the net output of kWh for a certain WECS design on a certain location.

L. L. FRERIS, IMPERIAL COLLEGE, LONDON, U.K.

When the authors state that the 4000 MW maximum infeed into their system is an excessive percentage - with respect to regulation and control purposes - of the 9000 MW maximum demand at winter, have they taken into account the interconnections of the Dutch system to neighbouring systems? This surely enhances the capacity of the Dutch system beyond the 9000 MW stated.

Authors' reply

We have not taken into account the interconnections of the Dutch system to the neighbouring systems, because these interconnections are certainly not meant to have your neighbours solve the problems which you cannot cope with yourself. Technically speaking the international interconnections could enhance the "absorption capacity", but only to a limited amount.

L.L. FRERIS, IMPERIAL COLLEGE, LONDON, U.K.

On the basis that there is a considerable correlation of wind activity over the country and electricity demand, WECS surely should be credited with some firm capacity. Would the authors comment?

Authors' reply

There is no doubt a rough correlation between wind activity and electricity consumption, but that does not help very much. It is the momentary correlation which counts. We have for instance not been able to discover a correlation between the daily load curve and wind speed pattern. The typical weather conditions during days on which maximum demand occurs could be described with: no wind, foggy, low temperature and dark.

R.J. LEICESTER, CENTRAL ELECTRICITY GENERATING BOARD, U.K.

I comment on the point raised by Mr. Freris that since wind contributes to electrical and other energy demands wind power can be considered to be firm power.

Indeed, the CEGB use wind forecasts and measurements to estimate demand. The relationship between wind speed and demand is complex and non-linear, however a simple approximation indicates that at low-moderate winds an increase of wind speed from 4 to 9 knots will give an increase in demand of about 1% on a typical winter week day. i.e. about 300 MW for the whole of England and Wales. At higher wind speeds saturation occurs and less increase in demand is experienced for the same 5 knot increase in wind speed.

The conclusion is therefore that although some wind power could perhaps be considered as firm on a statistical basis, the amount is insignificant in terms of a large scale system of wind generation.

Authors' reply

We would agree with this comment although we have not made investigations into this subject. We have the feeling that this sort of correlation is too weak generally speaking. The same applies to the seasonal correlation: there is more wind on the average and more electricity is consumed during the winter season. However, this sort of correlation does not help very much: it is the total lack of any correlation between the actual momentary power from WECS and the momentary load of the system which counts, in our opinion.

O. LJUNGSTROM, AERONAUTICAL RESEARCH INSTITUTE OF SWEDEN

I have a minor direct question of current interest for offshore WECS systems. Have you considered the pros and cons of the DC conversion line in offshore application?

Authors' reply

Not yet, but it will be studied in connection with a further study on off-shore systems in the framework of the national wind-energy program.

J. SMIT, NETHERLANDS ENERGY RESEARCH FOUNDATION

Which frequency intervals and voltage intervals are acceptable for WECS coupled to the national grid?

Are those intervals dependent on the number of WECS coupled to the grid?

Authors' reply

Frequency and voltage intervals can be defined for a system as a whole, not for a particular machine. As soon as a wind generator is connected to the grid the frequency is determined by the grid (if the grid is strong enough). The WECS must be kept in synchronism by means of the synchronising power of the grid.

L.L. FRERIS, IMPERIAL COLLEGE, LONDON, U.K.

With respect to the possible use of d.c. transmission, referred to in the paper, I would like to point out some further advantages. The series connection of windmills on the d.c. side requires only one cable and in the case of a cluster of Darrieus rotors it could be an overhead cable linking the tope of the adjacent windmill towers. As the ratio of cost of underground to overhead lines when towers are included is of the order or 10, it is likely that the suggested solution could result in a cost advantage of 20 or more.

If controlled rectifiers are used at the windmill converters, the fast control action inherent is a d.c. scheme can be arranged to contribute no significant short circuit capacity at the d.c. system connection.

Finally, I feel that the authors did not stress that an advantage of a d.c. compared to an a.c. system is that the former is an asynchronous link, therefore the windmills could be run at variable speed, i.e. at constant tip speed ratio, with a resulting larger energy yield per year.

Authors' reply

In our country more than 90% of the low voltage grid and more than 99% of the 10 kV network is underground. We do not think that people would accept overhead lines again for these lower voltages. The cost of these interconnections is of minor interest. We agree with the statements on short-circuit capacity and on the "asynchronous link".

R.J. LEICESTER, CENTRAL ELECTRICITY GENERATING BOARD, U.K.

I would like to offer some sympathy for the problems experienced by Mr. Bontius and his colleagues in their attempts to examine the integration of wind generators into an electricity supply system. In the U.K. we have looked at the problems in broad terms and there appear to be no particularly difficult problems until quite large amounts of wind generation (in excess of 20% installed peak capacity) are involved. However, the questions cannot be fully answered until we know the detailed characteristics of the output of large 1 MW and multi-megawatt machines, both individually and in large arrays. There are plenty of predictions of characteristics of all sorts of machines, but these are no substitute for genuine outputs and operational experience of large machines.

Since it would seem likely that Mr. Divone is going to be the first person to obtain such information, I would suggest that we request him to make the information available to all interested parties as soon as possible if this is allowed by the US ERDA programme.

INTRODUCTION OF LARGE SCALE WINDPOWER INTO AN EXISTING ELECTRICITY SUPPLY SYSTEM

G.H. Bontius

NV KEMA, Netherlands

If wind energy is supposed to cover some percent of the national energy balance, it should produce some 15% of the electricity production, since electric energy requires 20% of the primary fuels. If we take the net production of the public supply system of 60 TWh, wind energy should contribute 9 TWh (= 15%). According to present-day estimates this amount of electric energy can be produced by approximately 5000 wind turbines of 50m rotor diameter, having a power output of 1 MW at 12-13 m/sec. The supply system has a maximum load of 10,000 MW and a minimum load of 3500 MW. During large parts of the year it varies between 4000 MW and 9000MW. This means that many times a wind power has to be absorbed by the system which is of the same magnitude or even more than the actual load of the system. This applies to almost any electric supply system in the industrial countries.

In future utilities will have to cope with an increasing amount of non-controllable power, such as the power produced - and other combined H/P systems. The margin for a smooth interaction between wind turbines and the conventional system becomes less.

The power of 1 MW produced by one single unit at wind speeds of 12-13 m/sec or above is too large to be injected in the normal low and medium voltage (10 kV) distribution network. For bigger 10 kV sub-stations there is a possibility to connect some 10 or 20 units. In all circumstances a common network for groups of windturbines seems to be necessary which feeds into grids of at least 50-60 kV.

Unfortunately the situation arises that in areas where locations for WECS might be found the distribution network is weak (rural areas).

Discussion & Contributions

SESSION H : APPLICATIONS – SMALL SCALE

Chairman: P.J.Musgrove, Reading University, U.K.

Papers:

H1   An international development programme to produce a wind-powered water-pumping system suitable for small-scale economic manufacture
P.L.Fraenkel, Intermediate Technology Development Group Ltd., U.K.

H2   Wind energy-heat generation
R.Matzen, Royal Veterinary and Agricultural University, Denmark.

Contributions:

HX1   Good performance is important in Third World wind applications.
A.J.Garside, Cranfield Institute of Technology, U.K.

HX2   The economy of wind energy for glasshouse heating and other small scale applications.
C.Draijer, Consultant, Noorduijk, Netherlands.

Note:

The papers were presented by the authors whose names appear in bold print.

PAPER H1

Page H1-15   Replace Figures 3 and 4 by the Figures given below.

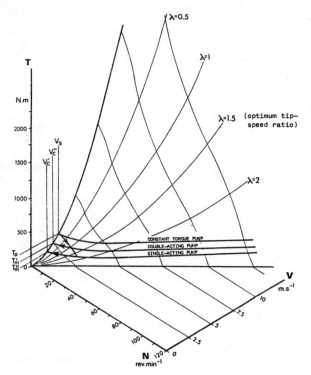

Operating lines for three different types of pump are indicated, for matching such that the windpump would start at wind velocity $V_S \simeq 3 ms^{-1}$ Velocities V_C' and V_C'' represent minimum velocities necessary for sustaining operation and produce the mean torque requirements for single- and double-acting pumps, shown as T_m' and T_m'' respectively. T_p is the peak torque requirement to start.

Fig. 3   Operating surface, showing torque (T) versus rotational speed (N) and wind velocity (V) for 6m diameter windpump with 24 blades giving rotor solidity of 82% measured at 0.7R.

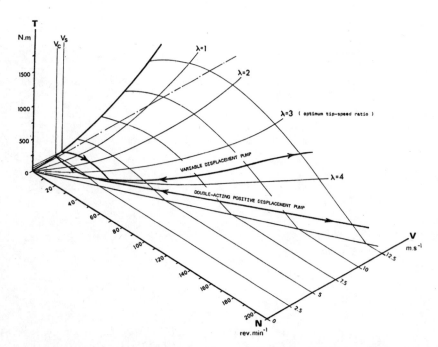

The two operating lines show the shortcomings of a positive displacement pump of fixed capacity in comparison with one of variable displacement. The hysteresis effect due to the different torque requirement for starting and sustaining motion with a double-acting pump can also be seen.

Fig. 4   Operating surface showing torque (T) versus rotational speed (N) and wind velocity (V) for a 6m diameter windpump fitted with 6 blades giving a rotor solidity of 14% as measured at 0.7R.

DISCUSSION

PAPER H1

An international development programme to produce a wind-powered water-pumping system suitable for small-scale economic manufacture

P. L. Fraenkel

Intermediate Technology Development Group Ltd., U.K.

F.C.EVANS, UNIVERSITY OF ST. ANDREWS, U.K.

Have you considered the use of peristaltic pumps, in which the torque characteristics might be well matched to those of the windmill and which could easily be changed from high-lift low-output to the opposite by changing the diameter of a piece of tubing?

Author's reply

Yes we did consider the use of peristaltic pumps, but there are major problems as far as I know in engineering a peristaltic pump capable of handling flow rates running into thousands of litres per hour since largish tubes that would be required for this will not readily stand up to being deformed. Also, I believe it would be difficult to operate a peristaltic pump with any degree of suction lift and it would not lend itself to immersion in a well and would certainly not easily fit down a bore hole. Also, I am not sure even if these objections could be overcome, that a peristaltic pump would offer a particularly good characteristic for matching a windmill rotor.

PAPER H2

Wind energy-heat generation

R. Matzen

Royal Vetinary and Agricultural University, Denmark

R.L.BALLARD, PENDAR TECHNICAL ASSOCIATES LTD., U.K.

We certainly should not under-estimate the need for heating in agriculture. In the U.K. a very large percentage of oil used in agriculture is used for heating greenhouses. Heat is relatively easy to store and it is natural to consider heating by using the wind. My company is giving some attention to ways of combining wind turbines with heat pumps because of the potential for increasing the heating capacity of a given size wind turbine. The use of heat pumps is also important where heat recovery is possible or where there is a simultaneous cooling requirement. I wonder if Dr Matzen has given any consideration to the use of heat pumps with the wind turbine and if he is carrying out any work in this field I can understand the need for simplicity in the heating system but I believe that this should not, by any means stop us from taking further an investigation of applications and economics.

Author's reply

The temperature level of the circulated water in the brakes heat exchanger may be chosen within a wide range up to 70-80°C. If higher temperatures are desired, the same system with oil as brake and heat exchanger media can be used, and the temperature level increases to 150°C.

I cannot see any practical idea in using heat pump to increase the temperature after the water brake.

On the other hand, if you think of using the wind turbine's mechanical power to run the compressor in a heat pump system, e.g. for the use in ventilation air from piggeries, ambient air or soil etc., you may probably lose the good power consumption v.s. rotational speed relationship, so you need an external r.p.m. control system on the wind turbines.

According to my knowledge of heat pumps, I do not think that the efficiency increases if the compressor runs with variable speed. -Concluding that wind turbine and heat pump systems with or without water-brake would not give a mutual efficiency increase.

B.SØRENSEN, NIELS BOHR INSTITUTE, DENMARK

If the heat load - say at a farm is displaced from the windy site, there would be an additional cost of heat transmission, which may exceed the cost of electricity transmission.

In the economic comparison, the cost of the water brake heat generator should be compared to that of a variable speed electric generator, which can work at the aerodynamical optimum of the wind turbine all the time. If the distance between wind turbine and heat load exceeds a few tens of meters, I believe that the conversion via electricity will be the most viable one.

Author's reply

A variable-speed electric generator, capable to work at the aerodynamical optimum of the wind, may not be less expensive in manufacture than a conventional asyncronous 3-phase generator connected directly on the AC-grid.

A water-brake may be produced, let us say, for 1 tenth of an electric generator of equivalent power rating.

The heat loss in insulated water pipes from the water-brake to the heat load can easily be calculated by traditional methods. I think your limitations of a few tens of meters are too pessimistic. Heat losses limit the use of the system, but losses over even a few hundred meters are acceptable.

You also have loss in electric systems, beginning in the generator with ca. 20%, but the main advantage of the water-brake and wind turbine systems is their simplicity in manufacture and maintenance.

M. LODGE, INSTITUTE OF MAN AND RESOURCES, CANADA

Dwelling on the fact that the ultimate user of the direct heat generation is a farmer who is familiar with the maintenance and repair of such mechanical equipment then it is apparent that such an approach is very appropriate. In the matter of costs it is probable that, because of the use of less expensive materials (i.e. very little copper) and perhaps less total material and little or no control equipment, the ultimate costs will be less.

Author's reply

Thank you for your comment. I agree that general maintenance and repair of the system is similiar to that of farmers' mechanical equipment.

C.DRAIJER, CONSULTANT, NOORDWIJK, NETHERLANDS

Professor Sørensen asked: why produce heat by means of a water brake instead of generating electricity.

The fact that the torque-angular speed characteristic of the windmill and a water brake are equal for every wind speed, means that the combination works always at maximum efficiency, up to the maximum angular speed for which the windmill is designed.

Taking into account the generator losses, network losses and the conversion losses to attain the final energy form, in this case heat, it shows that the yearly energy production per swept area unit of the water brake system is twice as high as the energy produced by an electric generator system.

Given the fact that the water brake system is cheaper than the electric generator together with the network and the energy converter, it is clear that the energy unit produced by the water brake system is much cheaper. In general the discussions about whether to use wind energy to generate the highly valuable electricity, or to generate the energy form which is needed on the spot are infact of not so much interest. In windy areas wind is everywhere and when for example heat is needed, it is of no use to consider further energy conversions when the conversion of the kinetic energy of wind into heat can be realised in competition with the burning of fossil fuels.

B.SØRENSEN, NIELS BOHR INSTITUTE, DENMARK

As stated in my comment, optimum C_p for every wind speed is not incompatible with an electric generator. I therefore still think, that even if heat generation equipment may be less expensive than the equivalent electricity generating equipment, than the savings may well be outbalanced by the higher cost of heat transmission than of electricity transmission plus secondary heat conversion (resistance heating or heat pump).

I expressed this as being a function of the distance of transmission, having seen many examples of farms, for which a windmill siting a few hundred meters away from the farm buildings would improve wind conditions dramatically (preferably also for safety reasons).

Author's reply

Mr Draijer's comment sounds well to my ears, and his statement, that running a windmill at constant tip speed ratio, compared to one with constant r.p.m. could give twice as much energy pr. swept area unit, is new to me.

It is evident, that the energy extraction from the wind is higher, but I have been most interested in the simple construction of the whole unit (propeller, transmission and generator).

GOOD PERFORMANCE IS IMPORTANT IN THIRD WORLD WIND APPLICATIONS

A.J.Garside

Cranfield Institute of Technology, U.K.

It is disappointing that only a small portion of the conference has been devoted to the application of wind power in the less developed countries. Perhaps we are concentrating too much to reduce our fuel bills by a few percent, when an equal effort should be devoted to providing a fuel bill for many people.

Our expertise must be applied to the large and small machines at the same time since the applications for cost and efficiency are the same, although cost may be measured in terms of kind rather than cash.

The cost can be at two levels (a) very low, suitable for purchase by an individual and thus be a simple machine, (b) rather higher, a machine that a community might buy for its own water needs having a high degree of reliability, and being more complex.

Efficiency is at a premium for all levels of complexity since, as Mr Fraenkel emphasizes the winds are generally only moderate, thus a machine of reasonable size and power output must have a high efficiency. This can be seen by plotting machine size for a given output against wind speed, for various efficiencies. Below winds of force 4 (5.5 m/s), lower efficiency machines require significant increases in size for the same power output.

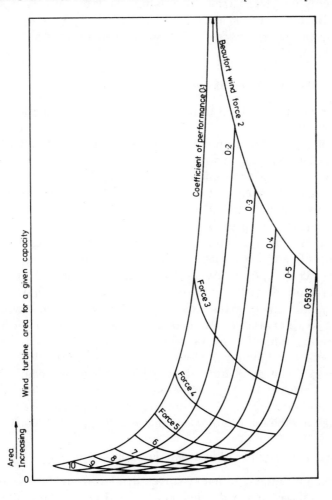

Fig. 1   The effect on a given capacity machine
of wind speed and coefficient of performance.

CONTRIBUTION HX2

THE ECONOMY OF WIND ENERGY FOR GLASSHOUSE HEATING AND OTHER SMALL SCALE APPLICATIONS

C.Draijer

Consultant, Noordwijk, Netherlands

1. Introduction

Characteristic for small scale applications is that the generated energy is used on the spot. That means that the economy is determined by the momentary demand. To evaluate the economy the differences between useful and generated energy must be known.

It is convenient to express this in a sort of efficiency factor, with ratio between wanted energy per year and total generated energy per year, what will be called the efficiency of use hereafter. It is obvious that this factor, depending on the specific application, is a function of the ratio of the capacity of the windmill installation to the total demand.

In close cooperation with the Dutch Institute of Agricultural Engineering, IMAG, in Wageningen, I made a study of the possible application of wind energy for the heating of glass houses.

In Western Holland about 5000 ha (12,500 acres) nett soil is heated under glass. Fuel is mainly natural gas, the use of which is stimulated by government in order to avoid air pollution in the densely populated area when using heavy oil, which was mainly in use before the natural gas well in Holland was found.

The amount of energy used for heating glasshouses is about 4% of the total energy consumption in Holland.

2. Determination of the efficiency of use

When generating heat the efficiency of use depends for glasshouses, as well as for any other application, on:
- the windenergy distribution as function of time, at a particular location and height.
- the demanded heat programme

The heating programme of a glasshouse depends on:
- the ambient air temperature distribution in time over the year.
- the solar radiation distribution
- the evaporation rate of the specific culture.
- the glasshouse inner temperature programme for a specific culture.

For relatively high heat rates as for example for lettuce, tomatoes and cucumbers (about 0,6 Gcal/m^{2}year or 6,7 10^3 kWh/m^{2}year),IMAG furnished the heating programmes. The windenergy distributions at the relative locations were determined from the available wind velocity distributions published by the Royal Dutch Meteorological Insitute, de Bilt.

Both data were available as month-average values over the year and as hourly-average values over a natural day specific for each month. From these data variation factors were gathered for the monthly and hourly periods, which multiplied gave the total variations of the generated and demanded energy quantities over the various periods. Up to 25% relative capacity of the windmill installation the efficiency of use showed to be 100%. At 40% relative capacity this factor was already down to about 90% and at 100% relative capacity (that means the installation is capable to produce all the energy needed on a yearly basis) the efficiency of use was 65%.

3. Application of wind energy for glasshouse heating

With a 100% efficiency of use a series produced medium size windmill for heat generation has the potentialities to become economical on the basis of the natural gas price for industrial use in Holland.

Fortunately there are various reasons not to try to reach a total covering of the energy demand by wind energy in glasshouse heating:
- a certain amount of carbondioxide production is needed for the culture
- the reliability of a glasshouse heating system must be very high, that means there is

always a redundancy in heating capacity, often divided over two or more units
- in the concentration areas of glasshouses a positioning of all the windmills needed, even to cover only 25% of the total demand, is not possible without significantly affecting their performance.

Although the application of windenergy in the field of agriculture seems to be disappointing at first sight, the relative portion will always be much more than the expected relative portion of wind- and solar energy together of the total energy consumption in Holland for the next twenty years.

4. Storage devices

The efficiency of use factor can be improved by applying means for storage of the momentary produced surplus energy. Because it is advantageous to let the windmill produce hot water by means of a waterbrake system, storage of hot water is a first choice. However storage in a convential hot water tank shows not to improve the economy. Other cheaper means for storage of the necessary quantities of hot water, more integrated with the glasshouse itself and the operation are subject of study now.

5. Other applications

The windenergy can furthermore be used in agriculture to produce:
- electricity for artificial lighting of glasshouses
- heat pump drive in such a heating system for glasshouses
- hot water for dairy farms

For all such applications the above holds more or less to the same extent. The more complicated and thus expensive the installation is the more the efficiency of use determines the economy.

6. Conclusion

In order to determine whether a small scale application of wind energy becomes economical the usefulness of the "installed" energy supply system must be determined for each particular application.

It will be found that only for a partial contribution wind energy will be economical.

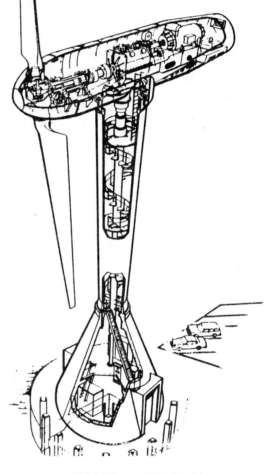

LIST OF EXHIBITORS

BHRA Fluid Engineering	Cranfield Bedford MK43 0AJ England	Tel: (0234) 750422 Telex: 825059	Contact: D. Leech
Elsevier Scientific Publishing Co.	Jan van Galenstraat 335 P. O. Box 330 1000 AH Amsterdam Netherlands	Tel: 020 5159222 Telex: 16479	Contact: Mr. Lambach
ERA Ltd.	Cleeve Road Leatherhead Surrey KT22 7SA England	Tel: Leatherhead 74151	Contact: R. Stafford
Holec Machines & Systemen	Postbus 50 Ridderkerk Netherlands	Tel: 01804-13633	Contact: J. Verwaal
NAKA	Sotriskestigen 10 611 00 Nykoping Sweden	Tel: 0155-12481	Contact: G. Ulfvengren
Real Gas and Electric Co. Inc.	278 Barham Avenue P. O. Box F Santa Rosa Ca. 95402 U.S.A.	Tel: 707-526-3400	Contact: S. Kagin
Saab-Scania AB and Stal-Laval	581 88 Linkoping Sweden	Tel: 46 13 129020 Telex: 50040	

WINDMILL INSTALLATIONS INDEX

(See also table on page A4-36 for installations in Japan).

Type	Institution Location etc.	Country	Rated Power /Diameter	Blade Number	Page Numbers
General	Rocky Flats Test Centre	U.S.A.	1 kW, 8kW, 40 kW		A3-18, A3-24
Horizontal Axis	ECN (Pelten)	Netherlands	300 kW/25m		A1-3, A1-6
	ERA/BADG/ NSHEB Consortium	U.K.	3-7MW/60m	2 fixed pitch	D3
	FDO-Engng. Consultants BV	Netherlands	25m		A1-5
	Futaoi-jima (JTTC)	Japan		3	A4-31, A4-38
	Gedser	Denmark	200 kW/24m	3	A3-22, D2
	Intermediate Technology Devp. Group	U.K.	6m	6	H1
	Japan Telephone & Telegraph Corp.(Oshima Island)	Japan	2kW/8m	2	A4-33, A4-43
(Adler type)	Kugoshima Agricultural College	Japan	4kW/6m	6	A4-30, A4-37
	Kalkugnen	Sweden	60kW/18m	2	B6-65, B6-67 B7-75, D1
	Kalkugnen	Sweden	MW size		A2-11
	Marine Safety Agency	Japan	3W/0.6m	3	A4-33, A4-42
	Marine Safety Agency	Japan	30W/2,25m	5	A4-33, A4-42
	NAI (Hydro-wind) Prince Edward Islands	Canada		3	B3-26
	NASA/ERDA Mod O (Plumbrook)	U.S.A.		2	A3-19, A3-22

Type	Institution Location etc.	Country	Rated Power /Diameter	Blade Number	Page Numbers
	Mod OA (Clayton; Culebra Island; Rhode Island)	Puerto Rico U.S.A.	200kW		A3-19, A3-25
	Mod 1 (Boone)	U.S.A.	2MW/200ft		A3-19
	Mod 2		$2\frac{1}{2}$MW/300ft	2	A3-19
	NV-101	Antarctica	2kW/1.2m	multi-bladed	A4-32, A4-41
	SIKOB/ Kockums (projected)	Sweden	7MW	2	B8-89, B8-107
	SIKOB/ Kockums Twin turbine	Sweden			B8-90 to B8-91, B8-107
	WTS Proto-type	Sweden	2-4MW/ 70-90m		B8-87
	Yamada's Up-Type	Japan	100W/1.8 m	2	A4-39, A4-31
			300W/2.4m	3	A4-40, A4-31, A4-32
			2kW	3	A4-33, A4-44
Shrouded Turbine (Horizontal axis)	Ben Gurion Univ. of the Negev	Israel	1kW/3m		F1
(inclined)	3-D-Triton (projected)	Sweden	20-30MW		B8-93
Darrieus (tropeskien)	Fokker-VFW	Netherlands	1.5m	2	A1-2
			5m	2	A1-2 to A1-3, A1-5, C3
	Japan Telephone & Telegraph Corp.		2kW/8m		A4-33
	Magdalen Islands	Canada	200kW/24m	2	E1, E2
	Poseidon I (projected)	Sweden	6.5MW/100m	2	B8-91 to B8-92, B8-108
	Poseidon II (projected)	Sweden	6.5MW	2	B8-92, B8-108

Type	Institution Location etc.	Country	Rated Power /Diameter	Blade Number	Page Numbers
	Poseidon III (projected)	Sweden			B8-92
	Sandia Labs. (Albuquerque)	U.S.A.	17m	2/3	A3-20, A3-26, E3
		projected	9kW/18ft 30kW/30ft 120kW/55ft 200kW/75ft 500kW/100ft 1600kW/150ft		E3
	Sanbagawa TV Relay Station	Japan	2kW/4m	3	A4-32, A4-42
	Vald turbine (projected)	Sweden	10-15MW		B8-93
Vertical Axis (straight bladed)	Flexible Blade Windmill	France		multi-bladed	E5
	Hacheimo-kojima	Japan	(7, 25Vd.c.)	multi-bladed	A4-31, A4-38
	Iwanaka Electric Works Ltd.	Japan	100kW, 1kW	4	A4-33, A4-44
(Cycloturbine)	Pinson Energy Corporation	U.S.A.	2-4kW/3.6m		E7
(Giromi 71)	Tokai Univ.	Japan	30W		A4-33, A4-44
			250W/2.5m	3	A4-33, A4-44
	Variable Geometry Vertical Axis Turbine (Reading University)	U.K.	3m	2	E4

SUBJECT INDEX

NAME INDEX

The index of authors and contributors includes the following:

    a) All authors of papers presented at the symposium.

    b) Those delegates who made contributions either verbally or in writing.

    c) Those delegates who made points and asked questions concerning the papers.

Matzen, R.	H2-17, Z17	Thornblad, P.	C6-89, Z32-Z33
	Z80-Z82	Tilenius, C. W. M.	Z23
Mays, I. D.	E4-39, Z33,	Tornkvist, G.	D1-1
	Z48-Z49	Tyndell, D. H.	D3-27, Z38-Z39
Mc Connell, R. D.	E2-11, Z45	Ushiyama, I.	A4-29, Z9-Z12
McKie, W. R.	E8-89		
Meggle, R.	Z72	Van den Berg, J. M.	B1-1, Z15-Z16
Meijer Drees, H.	E7-81, Z51	Van Essen, A. A.	B1-1, Z15-Z16
Milborrow, D. J.	Z18, Z62, Z63	Van Holten, T. H.	F2-13, Z62-Z65
Musgrove, P. J.	E4-39, Z9, Z23,	Van Rhijn, A. A. T.	x
	Z48-Z49, Z51,	Van Sant, J. H.	E2-11, Z45
	Z72, Z73	Vollan A.	C5-67, Z32, Z48
Nurick, G. N.	Z17	Warne, D. F.	iii, D3-27, Z38-Z39
		Watts, A.	E2-11, Z45
Ottens, H. H.	C3-31, Z31	Wendell, L. L.	B2-11
Pedersen, M.	iii	Wieringa, J.	Z15, Z20, Z24
Pelser, J.	Z11	Wilson, R. E.	E8-89
Pielke, R.	B4-33	Willmer, A. C.	Z52
Piepers, G. G.	iii, A1-1, B1-1,		
	Z4, Z15-Z16	Zwaan, R. J.	C3-31, Z31
Pontin, G. W. W.	Z21, Z22		
Rangi, R.	E1-1		
Rasmussen, F.	Z54		
Reichel, R.	A5-45, Z12		
Rosen, A.	C4-49, Z31-Z32		
Rothman, E. A.	C7-107		
Rowley, L.	Z8		
Sabzevari, A.	F3-25		
Sarre, P. E.	E9-101		
Schellens, F. J. C.	Z39		
Schmidt, W. L.	Z51		
Schulgasser, K.	F1-1, Z62		
Selzer, H.	iii, Z20, Z38		
Sens, P. F.	A1-1, Z4		
Sert, M.	Z3, Z12, Z18,		
	Z51		
Sforza, P. M.	Z66		
Shapiro, J.	iii, Z65, Z71		
Sharpe, D. J.	Z31, Z32		
Smedman	B7-73, Z22		
-Hogstrom, A. S.			
Smit, J.	Z74		
Smulders, P. T.	Z48		
Sockel, H.	Z15, Z30		
Sorensen, B.	G1-1, Z6, Z15		
	Z23, Z63, Z71-Z72		
	Z81, Z82		
South, P.	E1-1		
Stephens, H. S.	iii		
Stoop, T. H.	G3-27, Z73-Z75		
Templin, R. J.	iii, E1-1, Z6, Z17		
Ter Brugge, R.	B1-1, Z15-Z16		
Theyse, F. H.	Z33, Z48		

List of Delegates

ATTENDANCE LIST BY NAME OF DELEGATE

Abel-Aty E.	Chalmers T.U.	Sweden
Alestig R.	KMW AB	Sweden
Alexander A.J.	Loughborough Univ. of Technol	U.K.
Allen J.	E.T.S.U/A.E.R.E.	U.K.
Armstrong J.	Consultant	U.K.
Anderson M.B.	Cavendish Laboratory	U.K.
Antmarker M.	Mellby Industri AB	Sweden
Arnbak L.	Scangear Ltd	Denmark
Ballard R.L.	Pendar Technical Assoc. Ltd	U.K.
Banzhaf H.U.	Voith Getriebe KG	F.R.Germany
Bennenk M.H.	Van Hattum EN Blankevoort	Netherlands
Berneke E.	V.Kann Rasmussen Holdings	Denmark
Beyer R.	Meteorologie & Klimatologie Inst.	F.R.Germany
Bongaarts A.	Min. of Economic Affairs	Netherlands
Bontius G.H.	Kema N.V.	Netherlands
Braasch R.	Sandia Laboratories	U.S.A.
Brandels L.	Nat.Swedish Board for Energy Source Dev.	Sweden
Braun K.A.	Stuttgart Univ.	F.R.Germany
Builtjes P.J.H.	Organisation of Indust.Res.TNO	Netherlands
Byggeth N.	Statsforetag AB	Sweden
Calovolo M.	Fiat	Italy
Carlsson G.I.	Lutab	Sweden
Carstens C.	Fdo	Netherlands
Chamouton D.J.	Cytec	France
Chesneau D.V.	BP Research Centre	U.K.
Chezlepretre B.	Bertin & Cie.	France
Chikuni E.	Rural Industries Innovation	Botswana
Christensen C.J.	Riso National Laboratory	Denmark
Clayton B.R.	University College, London	U.K.
Coene R.	Delft Univ.of Technol.	Netherlands
Coppin P.A.	Meteorologie & Klimatologie Inst.	F.R.Germany
Crafoord C.	Stockholm University	Sweden
Dahlberg J.A.	Aeronautical Res.Inst.of Sweden	Sweden
Dalen G.V.	Fdo	Netherlands
De Lagarde J.M.	Montpellier University	France
De Vriefs D.	National Aerospace Lab.	Netherlands
Dekker J.W.M.	Netherlands Energy Res.Foundation	Netherlands
Delvig K.	Ministry of Environmental Prot.	Denmark
De Zeeuw W.J.	Eindhoven Univ.of Tech.	Netherlands
Divone L.V.	Dept. of Energy	U.S.A.
Doi A.	Hitachi Ltd.	Japan
Draijer C.	Consultant	Netherlands
Duggan J.G.	N.B.S.T.	Eire
Dumas P.	Dumas, Pierre, & Assoc.	Canada

Ekbom P.D.	Sydkraft AB	Sweden
Engstrom S.	Nat. Swedish Board for Energy Source Dev.	Sweden
Eriksson B.	Lund Inst. of Technol.	Sweden
Ethelfeld J.	Riso National Laboratory	Denmark
Evans F.C.	St. Andrew's University	U.K.
Fantom I.D.	BHRA	U.K.
Faxen T.	Uppsala University	Sweden
Fekete G.I.	McGill University	Canada
Fierens E.	Nobels Peelman	Belgium
Fraenkel P.L.	Intermediate Technol. Dev. Group	U.K.
Frandsen S.	Riso National Laboratory	Denmark
Freels J.	Berlin Tech. Univ.	F.R. Germany
Freris L.L.	Imperial Collge	U.K.
Friedman P.	California University	U.S.A.
Fritzsche A.	Dornier-System GMBH	F.R. Germany
Fuchs H.	Dornier-System GMBH	F.R. Germany
Gaffney F.	University College, Galway	Eire
Garside A.J.	Cranfield Inst. Technol.	U.K.
Gautier E.	Fiat	Italy
Gava P.	Eurocean	Monaco
Gherner L.	Gecu S.A.S.	Italy
Goodale B.A.	Kaman Aerospace Corp.	U.S.A.
Gratton M.Y.	Hydro Quebec	Canada
Graves J.	Boeing Eng & Constr.	U.S.A.
Gravesen S.	Denmark Tech. Univ.	Denmark
Gregoire R.P.	Hamilton Standard	U.S.A.
Griekspoor W.	Tebodin Consulting Eng.	Netherlands
Griffiths R.T.	Univesity College, Swansea	U.K.
Grylls W.M.	Exeter University	U.K.
Gustafsson A.	Aeronautical Res. Inst. of Sweden	Sweden
Hack R.K.	National Aerospace Lab.	Netherlands
Hagg F.	Fdo	Netherlands
Hardell R.	Sikob AB	Sweden
Hardy W.E.	Electrical Res. Association Ltd	U.K.
Hasbrouck	Hamilton Standard	U.S.A.
Hassan U.	Electrical Res. Association Ltd	U.K.
Hattestrand B.	VBB	Sweden
Hauekamp L.J.	Consultant	Netherlands
Helgesen H.	Lund University	Sweden
Hensing P.C.	Delft Univ. of Technol.	Netherlands
Herdin R.	Wiener Bruckenbau AG	Austria
Hinrichsen D.	"Ambia/New Scientist"	Sweden
Hiramoto A.	Fuji Electric Co. Ltd	Japan
Hirsch G.	Aachen T.U.	F.R. Germany
Hogstrom J.	Uppsala University	Sweden
Hollinger G.	Sunwind-Energy	Switzerland
Holme O.A.M.	Saab-Scania AB	Sweden
Hugosson S.	National Board Energy Source Dev.	Sweden
Huss G.	Man-Neue Technol	F.R. Germany

Igra D.	Ben Gurion University	Israel
Jackson R.O.	Midland Bank Ltd	U.K.
Jansen W.A.M.	Wind Energy Util.Proj.,Sri Lanka	Netherlands
Jarass A.	Regensburg Univ.	F.R.Germany
Jensen N.D.	Riso National Laboratory	Denmark
Jensen P.H.J.	O.V.E.	Denmark
Jespersen K.P.	Transmotor Ltd	Denmark
Johansson P.A.	Teleplan AB	Sweden
Kadlec F.G.	Sandia Laboratories	U.S.A.
Kagin S.	Real Gas & Electric Co.	U.S.A.
Kaine M.	Montreal University	Canada
Kern P.L.	Elektro GMBH	Switzerland
Ketley G.R.	British Aerospace Dynamics Group	U.K.
Kimura S.	T.T.I.	Japan
King P.S.	Environment Canada	Canada
Kvick T.	Swedish Met. & Hydro. Inst	Sweden
Lagerstrom B.L.	Teleplan AB	Sweden
Lamming S.D.	Reading University	U.K.
Lancel L.	Technip	France
Langer P.	Swedish State Power Board	Sweden
Larsson A.	Sydkraft AB	Sweden
Larsson L.	Lund Inst.of Technol.	Sweden
Le Gourieres D.	Dakar University	Senegal
Leicester R.J.	C.E.G.B.	U.K.
Leijonhufund S.	Teleplan AB	Sweden
Lembke J.	Danish Innovation centre	Denmark
Lindley D.	Taylor Woodrow Constr. Ltd	U.K.
Lippmann G.	Dornier System GMBH	F.R.Germany
Lissaman P.B.S.	AeroVironment Inc.	U.S.A.
Ljungstrom O.	Aeronautical Res.Inst.of Sweden	Sweden
Lodge M.A.	Inst.of Man & Resources	Canada
Lothigius J.	LUTAB	Sweden
Lowe R.J.	Open University	U.K.
Lundgren S.	Aeronautical Res.Inst.of Sweden	Sweden
Lundsager P.	Riso National Laboratory	Denmark
Luthander S.	LUTAB	Sweden
Lutz H.	Hoersch-Rote Erde Schmiedag	F.R.Germany
Magnussen E.	Atomenergi Inst.	Norway
Manders A.H.E.	Kema N.V.	Netherlands
Martinelli G.	Fiat	Italy
Masunaga N.	Toshiba Corporation	Japan
Matsunoshita S.	Nisshoelectric MFG. Co. Ltd.	Japan
Matzen R.	Roy.Vet. & Agr.Univ.	Denmark
Mays I.D.	Reading University	U.K.
Meggle R.	MBB Drehfluegler & Verkehr	F.R.Germany
Meijer Drees H.	Pinson Energy Corp.	U.S.A.
Metcalfe P.W.	Unarco-Rohn	U.S.A.
Mets V.	Swedish State Power Board	Sweden
Meyer R.	Regensburg University	F.R.Germany
Milborrow D.J.	C.E.G.B.	U.K.
Miura T.	IHI Co. Ltd	Japan
Molly J.P.	DFVLR	F.R.Germany

Morawetz E.	Teknoterm Systems AB	Sweden
Mowforth E.	"Wind Engineering"	U.K.
Musgrove P.J.	Reading University	U.K.
Naylor V.	Consultant	U.K.
Neumann R.	KFA Juelich	F.R.Germany
Nielsen P.	DEFU	Denmark
Nilsson J.E.V.	Chalmers Univ.of Technol	Sweden
Nitteberg M.R.	Atomenegi Inst.	Norway
Noel J.M.	Aerowatt Co.	France
Nurick G.N.	Cape Town University	South Africa
Offringa L.J.J.	Eindhoven T.H.	Netherlands
Olsson C.	Sikob AB	Sweden
Olsson G.	Kockums Shipyard	Sweden
Ory H.	Aachen T.U.	F.R.Germany
Ottens H.H.	National Aerospace Lab.	Netherlands
Pedersen H.B.	Denmark, Tech.Univ of.	Denmark
Pedersen T.F.	B & W-Turbo	Denmark
Pelser J.	Netherlands Energy Res.Foundation	Netherlands
Pentelow T.	Permali Ltd	U.K.
Pernpeinter R.	Man-Neue Technol	F.R.Germany
Petersen H.	Riso National Laboratory	Denmark
Piepers G.G.	Netherlands Energy Res.Foundation	Netherlands
Pignone G.A.	Fiat	Italy
Pontin G.W.	Wind Energy Supply Co. Ltd	U.K.
Pybus D.	Torrington Co. Ltd.	U.K.
Quraeshi S.	Shawinigan Eng.Co Ltd	Canada
Rasmussen F.	Riso National Laboratory	Denmark
Reichel R.	Dar Es Salaam University	Tanzania
Reid L.D.	Toronto University	Canada
Renaud P.W.	Ministry of Economic Affairs	Netherlands
Renner B.	Karlsruhe Univ.	F.R.Germany
Rowley L.P.	Canadair Ltd.	Canada
Schellens F.	FDO	Netherlands
Schliekelmann R.J.	Fokker-VFW	Netherlands
Schmidt B.	Saab-Scania AB	Sweden
Schmidt W.L.	Worldwind	Canada
Scott J.	Lyon Industrial Estates	Eire
Selzer H.	Erno Raumfahrttechnik	F.R.Germany
Sengupta G.	Niels Bohr Inst.	Denmark
Sens P.F.	Netherlands Energy Res.Foundation	Netherlands
Sert M.	Tubitak.Marmara Res.Inst.	Turkey
Sforza P.M.	New York Polytechnic	U.S.A.
Shapiro J.S.	Wind Energy Supply Co.	U.K.
Sharpe D.J.	Kingston Polytechnic	U.K.
Simpson P.B.	Taylor Woodrow Constr. Ltd.	U.K.
Smedman-Hogstrom A.	Uppsala University	Sweden
Smit J.	Netherlands Energy Res.Foundation	Netherlands
Smulders P.T.	Eindhoven T.U.	Netherlands
Snoek C.W.	Nova Scotia Tech.Coll.	Canada
Sockel H.	Wien Tech.University	Austria
Sørensen B.	Niels Bohr Inst.	Denmark

Soulez-Lariviere J.	S.F.E.D.	France
South P.	National Research Col.	Canada
Stephens H.S.	BHRA	U.K.
Stokes J.F.	Y-ARD Ltd	U.K.
Summerton J.	Nat. Swedish Board for Energy Source Dev.	Sweden
Sundsvold F.	Kvaerner Brug A/S	Norway
Svensson G.	Swedish State Power Board	Sweden
Swahn L.	ASEA AB	Sweden
Swart W.	Rhine Schelde Verolme	Netherlands
Taylor D.	"Architectural Design" magazine	U.K.
Templin R.J.	National Research Col.	Canada
Ter Brugge R.	Kema N.V.	Netherlands
Theyse F.H.	Theyse Energieberatung	F.R. Germany
Thoreson	Karlskruna Shipyard AB	Sweden
Thornblad P.G.	Stal-Laval Turbin AB	Sweden
Tilenius C.	Consultant	F.R. Germany
Tornkvist G.B.	SAAB-Scania AB	Sweden
Turkan E.	Netherlands Energy Res. Foundation	Netherlands
Turkcan E.	Netherlands Energy Res. Foundation	Netherlands
Ulfvengren	NAKA	Sweden
Ushiyama I.	Ashikaga Inst. of Technol.	Japan
Van Beek-Derwort M.	Delft Univ. of Technol.	Netherlands
Van Bussel G.J.W.	Delft Univ. of Technol.	Netherlands
Van Den Berg J.M.	Kema M.V.	Netherlands
Van Essen	Rykspanologische Dienst	Netherlands
Van Holton T.H.	Delft Univ. of Technol.	Netherlands
Van Hulle F.J.	A.T.O.L.	Belgium
Van Lonkhuyzen G.J.	"Hos-TV News"	Netherlands
Van'T Hul F.	T.H.T./S.W.D.	Netherlands
Vankjik G.A.M.	Delft Univ. of Technol.	Netherlands
Verhelst H.	Enka Research Lab.	Netherlands
Verwaal J.	Holec Machines & Systems	Netherlands
Vollan A.	Dornier-System GMBH	F.R. Germany
Watanabe I.	Matsushita Seiko Co. Ltd.	Japan
Warne D.F.	Electrical Res. Association Ltd	U.K.
Watts J.A.	Hydro Quebec	Canada
Weber W.	Voith Getriebe KG	F.R. Germany
Wendell L.L.	Battelle	U.S.A.
Wieringa J.	Royal Netherlands Met. Inst.	Netherlands
Wigstrand S.	Teleplan AB	Sweden
Wulff E.	Hamilton Standard	U.S.A.
Yahagi F.	Central Res. Inst of Electrical Power Indust.	Japan
Ykema T.	Kema N.V.	Netherlands
Zwaan R.J.	National Aerospace Lab.	Netherlands

ATTENDANCE LIST BY NAME OF COUNTRY

Austria	Wien Tech. University	Sockel H.
	Wiener Bruckenbau AG	Herdin R.
Belgium	A.T.O.L.	Van Hulle F.J.
	Nobels Peelman	Fierens E.
Botswana	Rural Industries Innovation	Chikuni E.
Canada	Canadiar Ltd.	Rowley L.P.
	Dumas, Pierre, & Assoc.	Dumas P.
	Environment Canada	King P.S.
	Hydro Quebec	Gratton M.Y.
		Watts J.A.
	Inst. of Man & Resources	Lodge M.A.
	McGill University	Fekete G.I.
	Montreal University	Kaine M.
	National Research Col.	South P.
		Templin R.J.
	Nova Scotta Tech. Coll.	Snoek C.W.
	Shawinigan Eng. Co. Ltd.	Quraeshi S.
	Toronto University	Reid L.D.
	Worldwind	Schmidt W.L.
Denmark	B & W-Turbo	Pedersen T.F.
	Danish Innovation Centre	Lembke J.
	Defu	Nielsen P.
	Denmark Tech. Univ.	Gravesen S.
	Ministry of Environmental Prot.	Delvig K.
	Niels Bohr Inst.	Sengupta G.
		Sørensen B.
	O.V.E.	Jensen P.H.J.
	Risø National Laboratory	Christensen C.J.
		Ethelfeld J.
		Frandsen S.
		Jensen N.D.
		Lundsager P.
		Petersen H.
		Rasmussen F.
	Roy. Vet. & Agr. Univ.	Matzen R.
	Scangear Ltd	Arnbak L.
	Transmotor Ltd	Jespersen K.P.
	V. Kann Rasmussen Holdings	Bernke E.
Eire	Lyon Industrial Estates	Scott J.
	N.B.S.T.	Duggan J.G.
	University College, Galway	Gaffney F.
Federal Republic of Germany	Aachen T.U.	Hirsch G.
		Ory H.
	Berlin Tech. Univ.	Freels J.
	Consultant	Tilenius C.
	Dornier System GMBH	Lippmann G.
		Fritzsche A.
		Fuchs H.
		Vollan A.

	DFVLR	Molly J.P.
	ERNO Raumfahrttechnik	Selzer H.
	Hoersch-Rote Erde Schmiedag	Lotz H.
	Karlsruhe Univ.	Renner B.
	KFA Juelich	Neumann R.
	Man-Neue Technol	Huss G.
		Pernpeinter R.
	MBB Drehfluegler & Verkehr	Meggle R.
	Meteorologie & Klimatologie Inst.	Beyer R.
		Coppin P.A.
	Regenburg University	Jaras A.
		Meyer R.
	Stuttgart Univ.	Braun K.A.
	Theyse Energieberatung	Theyse F.H.
	Voith Getriebe KG	Banzhaf H.U.
		Weber W.
France	Aerowatt Co.	Noel J.M.
	Bertin & Cie	Chezlepretre B.
	Cytec	Chamouton D.J.
	Montpellier University	De Lagarde J.M.
	S.F.E.P.	Soulez-Larivire J.
	Technip	Lancel L.
Israel	Ben Gurion University	Igra O.
Italy	Fiat	Calovolo M.
		Gautier E.
		Martinelli G.
		Pignone G.A.
	Geco S.A.S.	Gherner L.
Japan	Ashiwaga Inst. of Technol.	Ushiyama I.
	Central Res.Inst. of Electrical	
	Power Indust.	Yahagi F.
	Fuji Electric Co. Ltd.	Hiramoto A.
	Hitachi Ltd.	Doi A.
	IHI Co. Ltd.	Miura T.
	Matsushita Sieko Co. Ltd.	Watanabe I.
	Nisshoelectric MFG. Co. Ltd.	Matsunoshita S.
	Toshiba Corporation	Masunaga N.
	T.T.I.	Kimura S.
Monaco	Eurocean	Gava P.
Netherlands	Consultant	Drauer C.
		Hauerkamp L.J.
	Delft Univ.of Technol.	Coene R.
		Hensing P.C.
		Van Beek-Derwort M.
		Van Bussel G.J.W.
		Van Holton T.H.
		Vankuik G.A.M.
	Eindhoven T.H.	De Zeeuw W.J.
		Offringa L.J.J.
		Smulders P.T.

	Enka Research Lab.	Verhelst H.
	FDO	Carstens C.
		Dalen G.V.
		Hagg F.
		Schellens F.
	Fokker-VFW	Schliekelmann R.J.
	Holec Machines & Systems	Verwaal J.
	"Hos-TV News"	Van Lonkhuyzen G.J.
	Kemi N.V.	Bontius G.H.
		Manders A.H.E.
		Ter Brugge R.
		Van Den Berg J.M.
		Ykema T.
	Ministry of Economic Affairs	Bongaarts A.
		Renaud P.W.
	National Aerospace Lab.	De Vries O.
		Hack R.K.
		Ottens H.H.
		Zwaan R.J.
	Netherlands Energy Res. Foundation	Dekker J.W.M.
		Pelser J.
		Piepers G.G.
		Sens P.F.
		Smit J.
		Turkin E.
		Turkcan E.
	Organisation of Indust. Res. TNO	Builtjes P.J.H.
	Rhine Schelde Verolme	Swart W.
	Royal Netherlands Met. Inst.	Wieringa J.
	Rykspanologische Dienst	Van Essen
	Tebodin Consulting Eng.	Griekspoor W.
	T.H.T./S.W.D. ·	Van't Hul F.
	Van Hattum En Plankevoort	Bennenk M.H.
	Wind Energy Uttl. Proj., Sri Lanka	Jansen W.A.M.
Norway	Atomenergi Inst.	Magnussen E.
		Nitteberg M.R.
	Kvaerner Brug A/S	Sundsvold E.
Senegal	Dakar University	Le Gourieres D.
South Africa	Cape Town University	Nurick G.N.
Sweden	Aeronautical Res. Inst. of Sweden	Dahlberg J.A.
		Gustafsson A.
		Ljungstrøm O.
		Lundgren S.
	ASFA AB	Swahn L.
	Chalmers Univ. of Technol.	Abbel-Aty E.
		Nilsson J.E.V.
	Karlskrona Shipyard AB	Thoreson
	KMW AB	Alestig R.
	Kockums Shipyard	Olsson G.
	Lund Inst. of Technol.	Eriksson R.
		Larsson L.
	Lund University	Helgesen H.

	LUTAB	Carlsson G.I.
		Lothigius J.
		Luthander S.
	Mellby Industri AB	Antmarker M.
	Naka	Ulfvengren
	Nat.Swedish Board for Energy	
	Source Dev.	Brandels L.
		Engstrom S.
		Hugusson S.
		Summerton J.
	SAAB-Scania AB	Holme O.A.M.
		Schmidt B.
		Tornkvist G.B.
	SIKOB AB	Hardell R.
		Olsson C.
	Stal-Laval Turbin AB	Thornblad P.G.
	Statsforetag AB	Byggeth N.
	Stockholm University	Crafoord C.
	Swedish Met. & Hydro.Inst.	Kvick T.
	Swedish State Power Board	Langer P.
		Mets V.
		Svensson G.
	Sydkraft AB	Ekbom P.O.
		Larsson A.
	Teknoterm Systems AB	Murawetz E.
	Teleplan AB	Johansson P.A.
		Lagerstrom B.L.
		Leijonhufund S.
		Wigstrand S.
	Uppsala University	Faxen T.
		Hogstrom U.
		Smedman-Hogstrom A.
	VBB	Hattestrand B.
Switzerland	Elektro GMBH	Kern P.L.
	Sunwind-Energy	Hullinger G.
Tanzania	Dar Es Salaam University	Reichel R.
Turkey	Tubitak, Marmara Res.Inst.	Sert M.
U.K.	"Architectural Design" magazine	Taylor D.
	BHRA	Fantom I.D.
		Stephens H.S.
	BP Research Centre	Chesneau D.V.
	British Aerospace Dynamics Group	Ketley G.R.
	C.E.G.B.	Leicester R.J.
		Milborrow D.J.
	Cavendish Laboratory	Anderson M.B.
	Consultant	Armstrong J.
	Cranfield Inst.Technol.	Naylor V.
		Garside A.J.
	E.T.S.U/A.E.R.E.	Allen J.
	Electrical Res.Association Ltd	Hardy W.E.
		Hassan U.
		Warne D.F.

Exeter University	Grylls W.M.
Imperial College	Freris L.L.
Intermediate Technol. Dev. Group	Fraenkel P.L.
Kingtson Polytechnic	Sharpe D.J.
Loughborough Univ. of Technol.	Alexander A.J.
Midland Bank Ltd.	Jackson R.O.
Open University	Lowe R.J.
Pendar Technical Assoc. Ltd.	Ballard R.L.
Permali Ltd	Pentelow T.
Reading University	Lamming S.D.
	Mays I.D.
	Musgrove P.J.
St. Andrew's University	Evans F.C.
Taylor Woodrow Constr. Ltd	Lindley D.
	Simpson P.B.
Torrington Co. Ltd.	Pybus D.
University College. London	Clayton B.R.
University College. Swansea	Griffiths R.T.
Wind Energy Supply Co. Ltd.	Pontin G.W.
	Sharpiro J.S.
Y-ARD Ltd	Stokes J.E.
U.S.A. AeroVironment Inc.	Lissaman P.B.S.
Battelle	Wendell L.L.
Boeing Eng & Constr.	Graves J.
California University	Friedman P.
Dept. of Energy	Divone L.V.
Hamilton Standard	Gregoire R.P.
	Hasbrouck
	Wulff E.
Kaman Aerospace Corp.	Goodale B.A.
New York Polytechnic	Sforza P.M.
Pinson Energy Corp.	Meijer Drees H.
Real Gas & Electric Co.	Kagin S.
Sandia Laboratories	Braasch R.
	Kadlec E.G.
Unorco-Rohn	Metcalfe P.W.